Lectures in Mathematics
ETH Zürich
Department of Mathematics
Research Institute of Mathematics

Managing Editor:
Oscar E. Lanford

Anthony J. Tromba
Teichmüller Theory in Riemannian Geometry

based on lecture notes by
Jochen Denzler

2nd revised printing

Springer Basel AG

Author's address:

Anthony J. Tromba
Mathematisches Institut
Ludwig-Maximilians-Universität München
Theresienstr. 39
D-8000 München 2

Anthony J. Tromba
Department of Mathematics
University of California
Santa Cruz, CA 95064
USA

Deutsche Bibliothek Cataloging-in-Publication Data

Tromba, Anthony J.:
Teichmüller theory in Riemannian geometry: based on lecture notes by Jochen Denzler / Anthony
J. Tromba. – Springer Basel AG, 1992
(Lectures in mathematics)
ISBN 978-3-7643-2735-4   ISBN 978-3-0348-8613-0 (eBook)
DOI 10.1007/978-3-0348-8613-0

9 8 7 6 5 4 3 2

Preface

These lecture notes are based on the joint work of the author and Arthur Fischer on Teichmüller theory undertaken in the years 1980-1986. Since then many of our colleagues have encouraged us to publish our approach to the subject in a concise format, easily accessible to a broad mathematical audience.

However, it was the invitation by the faculty of the ETH Zürich to deliver the ETH Nachdiplom-Vorlesungen on this material which provided the opportunity for the author to develop our research papers into a format suitable for mathematicians with a modest background in differential geometry. We also hoped it would provide the basis for a graduate course stressing the application of fundamental ideas in geometry. For this opportunity the author wishes to thank Eduard Zehnder and Jürgen Moser, acting director and director of the Forschungsinstitut für Mathematik at the ETH, Gisbert Wüstholz, responsible for the Nachdiplom Vorlesungen and the entire ETH faculty for their support and warm hospitality.

This new approach to Teichmüller theory presented here was undertaken for two reasons. First, it was clear that the classical approach, using the theory of extremal quasi-conformal mappings (in this approach we completely avoid the use of quasi-conformal maps) was not easily applicable to the theory of minimal surfaces, a field of interest of the author over many years. Second, many other active mathematicians, who at various times needed some Teichmüller theory, have found the classical approach inaccessible to them.

The spirit of this approach was partially inspired by a paper of Earle and Eells "On a Fibre Bundle Description of Teichmüller theory" published in 1969 in the Journal of Differential Geometry, and is more in line with the traditional development of ideas in geometry and partial differential equations. Moreover we intended to have the material in this book, both the analytic as well as the geometric, reasonably self-contained.

Whereas various authors on classical Teichmüller theory omit fundamental analytical results like the existence and uniqueness of extremal quasi-conformal maps, we on the other hand include (although in an appendix) the existence and uniqueness of harmonic diffeomorphisms which form part of the analytical basis of this theory. We hope therefore that these notes will indeed find their intended broad audience.

There are many individuals who have contributed to the existing literature in classical Teichmüller theory. Unfortunately due to the limitations of the Nachdiplom-Vorlesungen we could not mention several of their important and interesting results, nor were we able to present a dictionary between our approach and the classical one. Some of these results are available in the books of Gardiner [40] and Lehto [65].

In these lectures on Teichmüller theory we develop the essentials of the subject from basic fundamentals with the main intention of making Teichmüller theory easy to learn. Our readers must remain the final judges as to whether we succeeded in this goal. We should also mention that there is some additional material included here that was not presented in the original lectures.

Several people have helped us substantially in our efforts. First the Sonderforschungsbereich 256 for Partial Differential Equations in Bonn under the direction of Stefan Hildebrandt and the Max-Planck-Institute under the direction of Friedrich Hirzebruch very generously supported the research which resulted in these notes. My thanks go to my friend, colleague, and co-author Arthur Fischer who taught me much of the advanced geometry I know.

Michael Buchner, Hans Duistermaat, Stefan Hildebrandt, Alan Huckleberry, Jerry Marsden, Dick Palais, Andrey Todorov, and Friedrich Tomi provided encouragement as well as mathematical inspirations. The proof of Poincaré's theorem, section 1.5, and of the collar lemma in the appendix are due to Tomi and the proof of the Mumford Compactness Theorem is due to Tomi and the author. Andreas Müller provided the details of the argument that $\mathcal{D}_0$ is contractible at the end of section 3.4. Kurt Strebel, Ralph Strebel, and Heiner Zieschang provided us with important historical information.

Our appreciation goes to our students and to Adimurthi, Horst Knörrer, Alfred Künzle, Serge Lang, Michael Struwe, Eugene Trubowitz, and Eduard Zehnder, who attended some or all of the lectures, and whose interest and comments added to the quality of our presentation. We owe a great thanks to Yair Minsky who found an error in our original approach to the Nielsen problem.

We also wish to thank Stefan Winiger for his careful typing of the manuscript and Artur Barczyk for professionally drawing the pictures. This book, however, could never have achieved its current polished form without the tireless and enthusiastic efforts of Jochen Denzler for which the author is deeply appreciative.

A.J. Tromba
ETH Zürich
June 1991

Contents

0 Mathematical Preliminaries

Let us collect some definitions and facts from differential geometry, which will be useful for our presentation:

Definition 0.1 A C^∞ n-manifold M (without boundary) is a paracompact topological Hausdorff space, together with a maximal collection of open subsets $(U_i)_{i \in I}$ covering M: $\bigcup_{i \in I} U_i = M$, and homeomorphisms $\varphi_i : U_i \to \mathbb{R}^n$ such that whenever $U_i \cap U_j \neq \emptyset$, $\varphi_i \circ \varphi_j^{-1}$ is C^∞. (The collection $\{(U_i, \varphi_i)\}$ has been suppressed in the notation here, as will be done, when no confusion can arise.)
M is said to be orientable, if the covering can be chosen so that $\det D(\varphi_i \circ \varphi_j^{-1})(\varphi_j(x))$ is always positive. If M is orientable, then, subject to this property, it has two possible maximal coverings. A choice of one is called an orientation.

A C^∞ Banach manifold M is a paracompact topological Hausdorff space, together with a maximal collection of open subsets $(U_i)_{i \in I}$, $\bigcup_{i \in I} U_i = M$, and of homeomorphisms $\varphi_i : U_i \to E$, where E is a Banach space, such that $\varphi_i \circ \varphi_j^{-1}$ is C^∞ whenever defined (i.e. whenever $U_i \cap U_j \neq \emptyset$).

If E is a Hilbert space, M is called a C^∞ Hilbert manifold.

Similarly, manifolds can be defined, which are modelled after any topological vector space E (say e.g. a Fréchet space), and which are of any differentiability class by stipulating that $\varphi_i \circ \varphi_j^{-1}$ should be of the corresponding class. We are especially interested in

Definition 0.2 A Riemann surface M is a C^∞ oriented manifold, together with a collection of local homeomorphisms $\varphi_i : U_i \to \mathbb{C}$ in the given orientation with $\bigcup U_i = M$ such that $\varphi_i \circ \varphi_j^{-1} : U_{ij} \to \mathbb{R}^2$ are holomorphic maps, where defined, i.e. if $\emptyset \neq U_{ij} := \varphi_j(U_i \cap U_j) \subset \mathbb{R}^2$.

Here, $\mathbb{R}^2$ is to be considered as $\mathbb{C}$. The collection $c := \{(U_i, \varphi_i) \mid i \in I\}$ is called a *complex structure* for M. If M has a fixed complex structure, we write (M, c) to denote M with its given complex structure.

Definition 0.3 A mapping $f : M \to N$ where $(M, c), (N, c')$ are Riemann surfaces is called holomorphic, iff for all charts $(V_j, \psi_j) \in c'$, $(U_i, \varphi_i) \in c$, the mapping $\psi_j \circ f \circ \varphi_i^{-1}$ is holomorphic (where defined). A similar definition holds for C^∞ maps between C^∞ manifolds. Note that the definition does not depend on the choice of the charts.

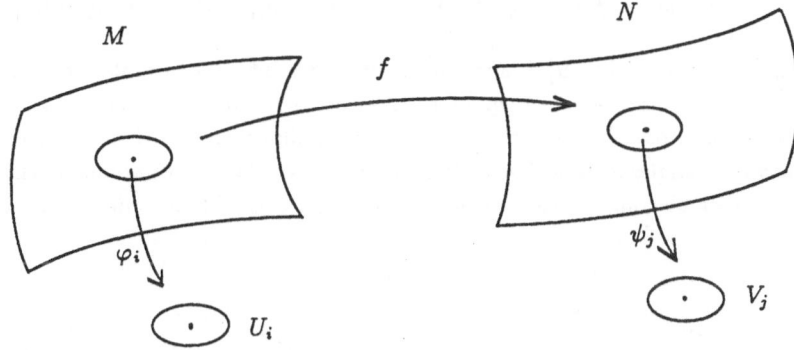

Figure 0.1: cf. definition 0.3

Definition 0.4

$f : M \to N$ is a diffeomorphism, iff $f \in C^\infty$ and f^{-1} is defined and $\in C^\infty$
$f : M \to N$ is a holomorphic equivalence, iff both f and f^{-1} are holomorphic.

A classical result in surface theory [43] yields

Theorem 0.1 *If M, N are compact, oriented C^∞ 2-manifolds (without boundary) which are homeomorphic, then they are diffeomorphic.*

This shows that the classification of compact oriented 2-manifolds without boundary up to diffeomorphisms is the same as their classification up to homeomorphism, so the

C^∞ structure brings no finer classification. The situation changes drastically for the complex structure. This was already known to Riemann:

Let c, c' be two complex structures on an oriented, compact 2-manifold M. Then the Riemann surfaces (M, c) and (M, c') need not be holomorphically equivalent. The existence of several distinct complex structures can be easily seen in the case of tori (genus 1). It is classical knowledge of course: A holomorphic equivalence $f : T = \mathbb{C}/\Omega \rightarrow T' = \mathbb{C}/\Omega'$ can be lifted to a holomorphic equivalence $\tilde{f} : \mathbb{C} \rightarrow \mathbb{C}$ which commutes with the translations by lattice vectors in Ω and Ω' respectively:

$$\tilde{f}(z + \omega_i) = \tilde{f}(z) + \omega_i' \quad \text{for basis vectors} \quad \omega_1, \omega_2 \in \Omega, \ \omega_1, \omega_2 \in \Omega' \ .$$

This implies $|\tilde{f}(z)| \leq A + B|z|$, and, using Cauchy's integral formula, holomorphy forces $\tilde{f}(z) = a + bz$. In other words Ω maps to Ω' under a similarity transformation. But there is a continuum of parallelograms, which are not similar to each other and give rise to tori which are not biholomorphically equivalent: in particular, two parallelograms P and P' with sides given by complex vectors ω_1, ω_2 and $\omega_1', \omega_2' \in \mathbb{C}$ will be similar iff $\omega_1/\omega_2 = \omega_1'/\omega_2'$.

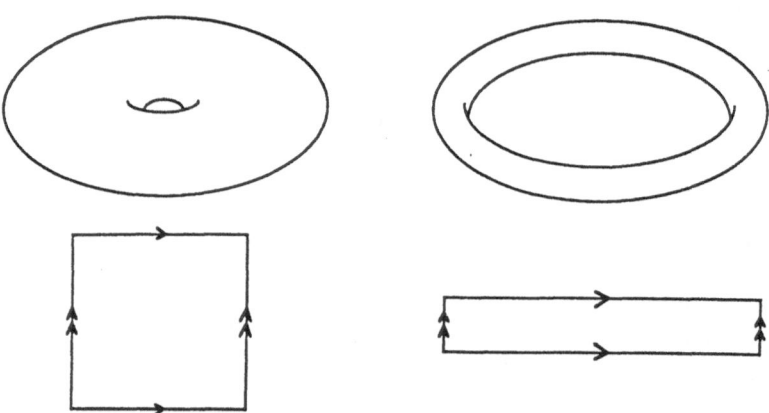

Figure 0.2: Tori and their corresponding parallelograms

As an aside, we note that the question of holomorphic (conformal) equivalence of Riemann surfaces is also equivalent to the question of isometry (up to scaling by a constant) of the corresponding Riemannian (real) 2-manifolds. In the case of tori these are all equipped with the "flat" metric (which comes from Euclidean $\mathbb{R}^2$); all are locally, but not in general globally isometric.

An analogous though less transparent situation holds for genus ≥ 2. Here again, holomorphic equivalence classes can be viewed as equivalence classes of isometries, but this time of hyperbolic metrics (metrics of constant negative curvature) as opposed to flat metrics (metrics of zero curvature) as in the case of a torus. We can argue this point heuristically as follows. The universal conformal covering space of a surface with a hyperbolic metric will be the hyperbolic plane. As the Euclidean plane can be tiled in many different ways by 4-gons, the hyperbolic plane can be tiled in many different ways by $4g$-gons ($4g$ of them meeting at each corner). If $g \geq 2$, these tilings give rise to globally non-isometric surfaces of genus g which are however locally isometric to the hyperbolic plane, their universal cover. Since the automorphisms (= holomorphic equivalences) of the upper half plane $\mathbb{H}$, considered as a Riemann surface, are the same as the isometry transformations of $\mathbb{H}$, considered as a Riemannian 2-manifold with the hyperbolic metric, we obtain many different holomorphic equivalence classes of a Riemann surface.

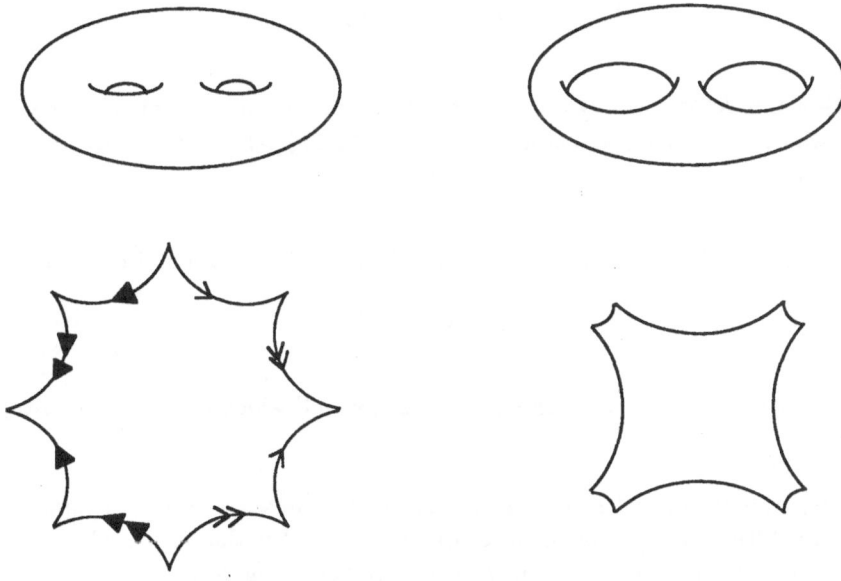

Figure 0.3: Surfaces of genus 2 and their corresponding octogons

These ideas will be carefully developed in the chapters that follow.

When classifying Riemann surfaces one does not attempt to distinguish *different* complex structures, but one distinguishes complex structures, which are *not holomorphically equivalent.*

If (M, c) is a Riemann surface and $f : M \to M$ a diffeomorphism, we can produce a new complex structure called the pullback of c by f. It is $f^*c := \left\{ (f^{-1}(U_i), \varphi_i \circ f) \right\}$ where $c = \{(U_i, \varphi_i)\}$. Now f^*c will be a different complex structure (different due to the charts, which are not only different from the former ones, but even incompatible to the former ones). Yet (M, f^*c) and (M, c) are holomorphically equivalent: $f : (M, f^*c) \to (M, c)$ is a holomorphic equivalence whereas, in general, $f : (M, c) \to (M, c)$ is not.

The classification of Riemann surfaces should be such that (M, c) and (M, f^*c) fall into the same equivalence class.

From now on, M will be fixed as a C^∞ 2-manifold and assumed to be of genus ≥ 2. Let $\mathcal{C}$ be the set of all complex structures c on M. Let $\mathcal{D}$ be the group of all C^∞ orientation preserving diffeomorphisms of M to itself. $\mathcal{D}$ acts on $\mathcal{C}$ (from the right) by pullback

$$
\begin{aligned}
\mathcal{C} \times \mathcal{D} &\to \mathcal{C} \\
(c, f) &\mapsto f^*c
\end{aligned}
$$

Definition 0.5 $\mathcal{R}(M) := \mathcal{C}/\mathcal{D}$ is called the Riemann moduli space of M. The quotient is taken with respect to the action just mentioned. $\mathcal{R}(M)$ will be the "space" of Riemann surfaces which are not holomorphically equivalent.

Riemann discovered by heuristic arguments [93], that it is a space of dimension

$$
6 \operatorname{genus}(M) - 6 \ .
$$

Up to now, we have not even considered a topology on $\mathcal{C}$ which would pass to a topology on $\mathcal{R}$.

A real breakthrough in understanding the equivalence classes of Riemann surfaces came, when Oswald Teichmüller thought to consider $\mathcal{C}/\mathcal{D}_0$ rather than $\mathcal{C}/\mathcal{D}$, where $\mathcal{D}_0$ is the subgroup of those diffeomorphisms in $\mathcal{D}$ which are homotopic to the identity.

Definition 0.6 $\mathcal{T}(M) := \mathcal{C}/\mathcal{D}_0$ is called the Teichmüller moduli space of M.

Teichmüller showed that it is a topological space homeomorphic to $\mathbb{R}^{6 \operatorname{genus}(M) - 6}$. It turns out that it is also a smooth finite dimensional manifold diffeomorphic to $\mathbb{R}^{6 \operatorname{genus}(M) - 6}$. The main goal of these lectures is to prove this last result and then go on to study the metric and complex structures on $\mathcal{T}(M)$.

Remark 0.1 *The reader might wonder about the relation between $T(M)$ and $\mathcal{R}(M)$. If a diffeomorphism homotopic to the identity is conjugated by any diffeomorphism, the result is again homotopic to the identity, so $\mathcal{D}_0$ is a normal subgroup of $\mathcal{D}$.*
By a fundamental theorem of Nielsen [81] and Baer [9], $\mathcal{D}/\mathcal{D}_0$ is isomorphic to $\mathrm{Aut}\,\pi_1(M)/\mathrm{inner}\mathrm{Aut}\,\pi_1(M)$, the group of outer automorphisms of the fundamental group of M, and this is therefore an infinite discrete group. Thus

$$\mathcal{R}(M) = \mathcal{C}/\mathcal{D} = (\mathcal{C}/\mathcal{D}_0)/(\mathcal{D}/\mathcal{D}_0) = T(M)/(\mathcal{D}/\mathcal{D}_0)$$

(to our present knowledge, at least as sets).
We will not go into a study of these types of questions here. The existence of a homomorphism from $\mathcal{D}/\mathcal{D}_0$ to $\mathrm{Aut}\,\pi_1(M)/\mathrm{inner}\mathrm{Aut}\,\pi_1(M)$ uses only general properties of the fundamental group functor. The basic reference for surjectivity is [81, Satz 11], for injectivity [9]. For more recent proofs see [72], [127], [98]. Further relevant information can be found in [131], [130], [51]. In principle, the quoted references prove the theorem for homeomorphisms instead of diffeomorphisms, but the proofs carry over.
The reader may, however, wish to see an example of an orientation preserving diffeomorphism which is not homotopic to the identity. Consider a non-trivial rotation of the fundamental domains in figures 0.2 and 0.3. In both cases, it induces a diffeomorphism of M which is not homotopic to the identity. For $\mathrm{genus}(M) = 1$, this diffeomorphism corresponds to the outer automorphism $x \mapsto -x$ of $\pi_1(M) \cong (\mathbb{Z}^2, +)$. For $\mathrm{genus}(M) = 2$, it corresponds to the outer automorphism of $\pi_1(M) \cong \langle a, b, c, d \mid aba^{-1}b^{-1}cdc^{-1}d^{-1} \rangle$ which is given by $a \mapsto c$, $c \mapsto a$, $b \mapsto d$, $d \mapsto b$.

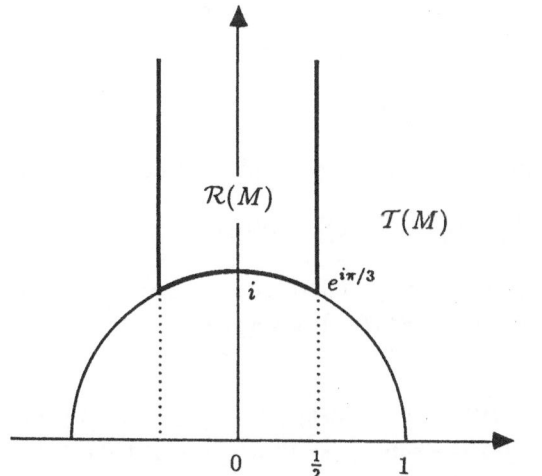

The reader who is familiar with elliptic functions is encouraged to compare the situation claimed here with the analogous situation in genus 1: $T(M)$ is the upper half plane, $\mathcal{R}(M)$ is the module figure, $\mathcal{D}/\mathcal{D}_0$ the modular group, the group of base changes in the lattice corresponding to M. The theory to be developed here will, however, not depend on knowledge of elliptic functions.

Figure 0.4: The analog to Teichmüller space for genus 1

Let $H^s(M) = H^s(M, \mathbb{R})$ be the linear space of all (real valued) functions on M whose derivatives up to order s exist in the sense of distributions and are square integrable functions. We say, these functions belong to the Sobolev class H^s.

Let $\mathcal{H}^s(T_p^q M)$ be the linear space of all (real valued) p-q-tensors on M which belong to the Sobolev-class H^s, i.e. functions $x \mapsto f(x)$ with $x \in M$ and $f(x) \in (T_x M)^{\otimes p} \otimes (T_x^* M)^{\otimes q}$, whose derivatives up to order s (which exist in the sense of distributions) are square integrable. For the definition of Sobolev spaces, see e.g. [62], [75], [3].

We will mainly be concerned with 1-1 tensors and 0-2 tensors. Therefore, let us give expanded versions of the definitions in these cases.

Definition 0.7 A symmetric 0-2 tensor h is a mapping which assigns to each x a symmetric bilinear form $h(x) : T_x M \to \mathbb{R}$. Given a differentiable map $\psi : U \to M$, one can pull back a 0-2 tensor on M to U:

$$(\psi^* h)(y)(\xi, \eta) := h\left(\psi(y)\right)\left(D\psi(y)\xi, D\psi(y)h\right) \ .$$

If ψ is the inverse mapping of a coordinate chart, this definition gives the pullback of a 0-2 tensor on M to a region in $\mathbb{R}^n$, where it is represented by a symmetric matrix at each point. h is called an H^s tensor, if the map $y \mapsto (\psi^* h)(y)$ is H^s as a map of U into the symmetric matrices where ψ^{-1} is a coordinate chart. Similarly, C^∞ tensors are defined.

Definition 0.8 A 1-1 tensor H is a mapping which assigns to each x a linear map

$$H(x) : T_x M \to T_x M \ .$$

Similarly as above, the pullback of H with respect to a diffeomorphism ψ can be defined by

$$(\psi^* H)(y) := D\psi^{-1}\left(\psi(y)\right) \circ H\left(\psi(y)\right) \circ D\psi(y) \ .$$

Therefore, in coordinates, a 1-1 tensor is represented by a matrix at each point. It is H^s (C^∞), if the map associating this matrix to each point is H^s (C^∞) as a map of U into the matrices.

Along with the spaces $\mathcal{H}^s(T_1^1 M)$ and $\mathcal{H}^s(T_0^2 M)$, we define the spaces $C^\infty(T_1^1 M)$ and $C^\infty(T_0^2 M)$ of C^∞ tensors of type 1-1 or 0-2 respectively and the spaces $S_2^s(M)$, $S_2(M)$ of symmetric 0-2 tensors on M of class H^s or C^∞ respectively. Since we consider M fixed, we will usually write S_2^s, S_2 for $S_2^s(M), S_2(M)$.

In dimension 2, Sobolev's embedding theorem guarantees that H^s functions (tensors) are C^0 if $s > 1$. We shall need one important consequence of Sobolev's embedding theorem: if $s > 1$ and f and g are of Sobolev class H^s, then so is $f \cdot g$, and hence H^s functions form an algebra.

We will assume $s > 3$ throughout the exposition to assure the algebra property for the derivatives of functions we consider as well as for the functions themselves.

1 The Manifolds of Teichmüller Theory

1.1 The Manifolds $\mathcal{A}$ and $\mathcal{A}^s$

Definition 1.1.1 The space of H^s - almost complex structures is defined to be

$$\mathcal{A}^s := \left\{ J \in \mathcal{H}^s(T_1^1 M) \mid \forall x : J_x^2 = -id_x; \right.$$
$$\left. \forall X_x \in T_x M : (X_x, J_x X_x) \text{ is an } oriented \text{ basis for } T_x M \right\} .$$

The space of C^∞ almost complex structures, $\mathcal{A}$, is defined similarly with $\mathcal{A} = \bigcap_s \mathcal{A}^s$.

The second axiom is used to distinguish J from $-J$ (which corresponds to distinguishing $-i$ from $+i$ in $\mathbb{C}$) by means of orientation.

On the standard $\mathbb{R}^2$, one might take $\hat{J} := \begin{pmatrix} 0 & -1 \\ 1 & 0 \end{pmatrix}$, which corresponds to multiplication by i on $\mathbb{C}$. This correspondence between the almost complex structures on $\mathbb{R}^2$ and the complex structures on $\mathbb{C}$ can be transferred to any Riemannian 2-manifold by means of local coordinate charts.

Namely, define for any coordinate chart $\varphi : U \to \mathbb{C}, (U, \varphi) \in c$

$$J_\varphi(x) := d\varphi_x^{-1} \hat{J} d\varphi_x . \qquad (1.1)$$

That this defines an almost complex structure on U is obvious; the following calculation will show that J_φ does not depend on φ, no matter which $\varphi \in c$ is taken. Therefore (1.1) associates an almost complex structure $\Gamma c \in \mathcal{A}$ to any complex structure $c \in \mathcal{C}$.

Let $(V, \psi) \in c$ be another chart, intersecting (U, φ). Then

$$J_{\psi}(x) = d\psi_x^{-1} \hat{J} d\psi_x \ , \ \ J_{\varphi}(x) = d\varphi_x^{-1} \hat{J} d\varphi_x \ .$$

Now,

$$J_{\varphi}(x) = J_{\psi}(x) \Leftrightarrow d(\varphi \circ \psi^{-1}) \hat{J} d(\psi \circ \varphi^{-1}) = \hat{J} \ . \tag{1.2}$$

But $\psi \circ \varphi^{-1}$ is a holomorphic map iff $d(\psi \circ \varphi^{-1})$ has the form $\begin{bmatrix} A & -B \\ B & A \end{bmatrix}$. An easy calculation shows immediately that this is equivalent to (1.2). So we have proved the following theorem, except for the surjectivity:

Theorem 1.1.1 *There exists a bijective mapping* $\Gamma : \mathcal{C} \to \mathcal{A}$. *It is defined in local coordinates by (1.1).*

The proof that Γ is surjective is postponed until after lemma 1.3.6.

Theorem 1.1.2 *The groups $\mathcal{D}$ and $\mathcal{D}_0$ act on $\mathcal{A}$ or $\mathcal{A}^{\bullet}$ by*

$$(f^* J)_x := (df_x)^{-1} J_{f(x)} df_x \quad \text{(for all } f \in \mathcal{D}) \ .$$

The map Γ is $\mathcal{D}$ equivariant, i.e.

$$\Gamma(f^* c) = f^* \Gamma(c) \ .$$

PROOF: Let $\varphi \in c$, $\varphi \circ f \in f^* c$. Then $d(\varphi \circ f)^{-1} \hat{J} d(\varphi \circ f) = df^{-1}(d\varphi^{-1} \hat{J} d\varphi) df$.
Now $J(f^* c)$ is the collection of the left hand side terms, $f^* J(c)$ is the collection of the right hand side terms. ∎

The last two theorems make it clear that it is enough to understand $\mathcal{A}/\mathcal{D}_0$ in order to understand $\mathcal{C}/\mathcal{D}_0$. All properties of $\mathcal{C}$, which is a set of charts, can be formulated with no loss of information as properties of $\mathcal{A}$, which is just a subset of the linear space of C^{∞} smooth 1-1 tensors, $C^{\infty}(T_1^1 M)$. If $\mathcal{A}$ were a C^{∞} Banach manifold and if $\mathcal{D}_0$ were a C^{∞} Banach Lie group, and if $\mathcal{D}_0$ acted smoothly, freely and properly on $\mathcal{A}$, then a standard result (see theorem 1.3.2 later) would guarantee that $\mathcal{A}/\mathcal{D}_0$ is a manifold. Such a result relies on the implicit function theorem (which holds, e.g., in Banach spaces). The difficulty is that we are working in the C^{∞} category, where no implicit function theorem is available (especially, since spaces of C^{∞} functions and tensors are not Banach spaces).

On the other hand, $\mathcal{D}_0^{\bullet}$ does not act on $\mathcal{A}^{\bullet}$, because the differential of a diffeomorphism enters in the definition of the action. What does act on $\mathcal{A}^{\bullet}$ is $\mathcal{D}_0^{\bullet+1}$, (i.e. the space of homeomorphisms of class $\mathcal{H}^{s+1}$ such that the inverse is of class $\mathcal{H}^{s+1}$, too). In this category, an implicit function theorem is available, but the action is not smooth. Indeed,

$$(f^*J)(x) = df_x^{-1} J_{f(x)} df_x \ .$$

So, $J \circ f$ enters, and this does not depend smoothly on f. The loss of one derivative in dJ when one tries to differentiate with respect to f leads one out of the class $\mathcal{H}^s$. A first goal in understanding $\mathcal{A}/\mathcal{D}_0$ is to understand $\mathcal{A}$. A step in this direction is

Theorem 1.1.3 $\mathcal{A}^s$ *is a* C^∞ *smooth submanifold of* $\mathcal{H}^s(T_1^1 M)$ *whose tangent space at a point* $J \in \mathcal{A}$ *is characterized by*

$$T_J \mathcal{A}^s = \{H \mid HJ = -JH\} \ .$$

PROOF: Before giving a correct proof, let us see, how a straightforward approach to a proof fails:

$$\Phi : \mathcal{H}^s(T_1^1 M) \to \mathcal{H}^s(T_1^1 M), \quad J \mapsto J^2$$

is a C^∞ mapping. $\mathcal{A}^s$ is the pre-image of $-I$. One would like to use the implicit function theorem.

$$D\Phi(J)H = HJ + JH$$

If $D\Phi(J)$ were surjective, this would prove that $\mathcal{A}^s$ is a C^∞ submanifold, the tangential space being $\ker D\Phi(J)$, which is just the formula claimed.

But, alas, $D\Phi(J)$ is not surjective for any $J \in \Phi^{-1}(-I)$ at any point. Every element $HJ + JH$ in its image commutes with J, whereas there are certainly tensors which do not commute with J.

A correct proof can be obtained as a consequence of the following facts, which will be established in a moment.

(1) For a 1-1 tensor J, it holds: $J^2 = -I \Leftrightarrow tr\, J = 0$ and $\det J = 1$

(2) $\mathcal{N} := tr^{-1}(0) \subset \mathcal{H}^s(T_1^1 M)$ is a linear subspace, and consequently a C^∞ submanifold with $T_J \mathcal{N} = \{H \mid tr\, H = 0\}$

(3) $\mathcal{M} := \det^{-1}(1) \subset \mathcal{H}^s(T_1^1 M)$ is a C^∞ submanifold with $T_J \mathcal{M} = \{H \mid tr\, JH = 0\}$

(4) If $\mathcal{M}, \mathcal{N}$ are C^∞ submanifolds of some Banach space $\mathcal{H}$ which intersect transversally, i.e.

$$\forall J \in \mathcal{M} \cap \mathcal{N} : T_J \mathcal{M} + T_J \mathcal{N} = \mathcal{H} \ ,$$

then $\mathcal{M} \cap \mathcal{N}$ is a C^∞ submanifold of $\mathcal{H}$, too, and $T_J(\mathcal{M} \cap \mathcal{N}) = T_J \mathcal{M} \cap T_J \mathcal{N}$.

(5) If $J^2 = -I$ then:

$$tr\, H = 0 \text{ and } tr JH = 0 \Leftrightarrow JH + HJ = 0 \ .$$

Due to (1), $\mathcal{A}^s = \mathcal{M} \cap \mathcal{N}$ as a set; (2) – (4) show that $\mathcal{A}^s$ is a C^∞ manifold, (5) brings the description of its tangent space back to the form given in the theorem.

Here are the verifications of the single steps:

ad (1) $J^2 - (tr\, J)J + (\det J) \cdot I = 0$ (for all 1-1 tensors J)
"$\Rightarrow$" is obvious from this equation.
If $J^2 = -I$ and $v \neq 0$, a vector field defined on a neighbourhood, then $v_x, J_x v_x$ is a basis for $T_x M$ with respect to which J has the form $J = \begin{pmatrix} 0 & -1 \\ 1 & 0 \end{pmatrix}$, hence $tr\, J = 0, \det J = 1$. (This is a local argument: v need not be a globally defined non-vanishing vector field, which for genus$(M) > 1$ does not exist.)

ad (2)   trivial

ad (3) $\frac{d}{dt}\big|_{t=0} \det(J + tH) = \det J \cdot \frac{d}{dt}\big|_{t=0} \det(I + tJ^{-1}H) = \det J \cdot \frac{d}{dt}\big|_{t=0} \det e^{tJ^{-1}H} =$
$\det J \cdot \frac{d}{dt}\big|_{t=0} e^{t \cdot tr\, J^{-1}H} = \det J \cdot tr\, J^{-1}H.$
Since $\det J = 1, J^{-1} = -J$, this is equal to $-tr\, JH$. It remains to show that $H \mapsto -tr\, JH$ is surjective. Then the implicit function theorem shows that $\mathcal{M}$ is a submanifold with tangent space $\ker(D \det(J)) = \{H \mid tr\, JH = 0\}$.
Indeed, given any $\rho \in H^s(M)$, choose $H := \frac{1}{2}\rho J \in \mathcal{H}^s(T_1^1 M)$, then $-tr(JH) = \rho$.

ad (4)   This is a general fact from differential topology which works as well for Banach manifolds. See, e.g. [17, (10.2.1)], [2], [63].

ad (5)   This is a point computation, so we can assume that H and J are 2×2 matrices. Let $\hat{J} = \begin{pmatrix} 0 & -1 \\ 1 & 0 \end{pmatrix}$. Since

$$tr\, H = 0 \Leftrightarrow tr\, S^{-1}HS = 0, \text{ and } tr\left((S\hat{J}S^{-1})H\right) = 0 \Leftrightarrow tr\left(\hat{J}(S^{-1}HS)\right) = 0,$$

it holds that $(S\hat{J}S^{-1})H + H(S\hat{J}S^{-1}) = 0 \Leftrightarrow \hat{J}(S^{-1}HS) + (S^{-1}HS)\hat{J} = 0$, and since any matrix J with $J^2 = -id$ can be written as $J = S\hat{J}S^{-1}$, it is enough to consider the case $J = \hat{J}$. This is a trivial calculation.   ■

1.2   The Riemannian Manifolds $\mathcal{M}$ and $\mathcal{M}^s$

Let $\mathcal{M}^s$ be the space of $\mathcal{H}^s$ Riemannian metrics on M, i.e.

$$\mathcal{M}^s := \left\{ g \in S_2^s \mid g(x)(u, u) > 0 \quad \text{if} \quad u \neq 0 \right\} .$$

$\mathcal{M}^s$ is an open subset of S_2^s (indeed, a cone) and thus a submanifold of S_2^s whose tangential space at any point can be naturally identified with S_2^s itself: $T_g\mathcal{M}^s \cong S_2^s$. Similarly, $\mathcal{M}$ is defined to be the space of all C^∞ Riemannian metrics. Let us also define $\mathcal{M}_{-1}$ and $\mathcal{M}^s_{-1}$ as the subsets of $\mathcal{M}$ and $\mathcal{M}^s$ which consist of the metrics with scalar curvature -1. They will become important in section 1.6.

$\mathcal{D}$ and $\mathcal{D}_0$ act on $\mathcal{M}^s$ and $\mathcal{M}$ by

$$(f^*g)(x)(u, v) := g(f(x))(Df(x)u, Df(x)v) .$$

There is a natural Riemannian metric on $\mathcal{M}^s$ itself, i.e. a symmetric positive definite bilinear form on S_2^s at every $g \in \mathcal{M}^s$:

Let $\xi, \eta \in T_g\mathcal{M}^s \cong S_2^s$.

g induces a scalar product on any space of tensors by "raising or lowering indices"; in our case:

$$(\xi \cdot \eta)(x) = g^{ik}(x)g^{lm}(x)\xi_{il}(x)\eta_{km}(x) \quad \text{in local coordinates.}$$

In invariant form, let $E, N \in \mathcal{H}^s(T_1^1 M)$ be defined by

$$\begin{aligned} \xi(x)(u_x, v_x) &= g(x)(E_x u_x, v_x) \\ \eta(x)(u_x, v_x) &= g(x)(N_x u_x, v_x) . \end{aligned}$$

Then $(\xi \cdot \eta)(x) = tr\, E_x N_x.$, an H^s function on M.

On the other hand, there is a natural volume element associated with g, namely

$$\mu_g(X_x, Y_x) := \pm \sqrt{\det \begin{bmatrix} g_x(X_x, X_x) & g_x(X_x, Y_x) \\ g_x(Y_x, X_x) & g_x(Y_x, Y_x) \end{bmatrix}} ,$$

where the $+$ sign is chosen if (X_x, Y_x) is a positively oriented base of T_xM, and the $-$ sign if it is negatively oriented.

After these definitions, the Riemannian metric on $\mathcal{M}^s$ is simply $\langle\langle \xi, \eta \rangle\rangle_g := \int_M \xi \cdot \eta\, d\mu_g$, where $d\mu_g$ denotes integration with respect to the Riemann volume measure induced by μ_g.

This metric is called the L^2-*metric*. It is well known in mathematical physics. $\mathcal{M}$ is equipped with the same metric.

The following decomposition theorem is of basic importance.

Lemma 1.2.1 *There is an L^2-orthogonal splitting $T_g\mathcal{M}^s = (S_2^s(g))^c \oplus (S_2^s(g))^T$, where:*

$$(S_2^s(g))^c := \left\{ h \mid h(x) = p(x) \cdot g(x), p \in H^s(M) \right\}$$
$$(S_2^s(g))^T := \left\{ h \mid tr_g H = 0 \right\}$$

(c stands for "conformal", T for "traceless").

PROOF:

$$h = \frac{1}{2}(tr_g h) \cdot g + \left(h - \frac{1}{2}(tr_g h) \cdot g \right)$$

is a decomposition corresponding to the given spaces, as one sees immediately. Since $tr_g(p \cdot g)(x) = 2p(x)$, the intersection $(S_2^s(g))^c \cap (S_2^s(g))^T = \{0\}$.

Orthogonality: Let h be traceless; then $pg \cdot h = pg^{ij}g^{kl}g_{ik}h_{jl} = p \cdot g^{jl}h_{jl} = 0$. ■

1.3   The Diffeomorphism $\mathcal{M}^s/\mathcal{P}^s \cong \mathcal{A}^s$

There are some basic results we will need for this section. The first of them is

Theorem 1.3.1 (Existence of conformal coordinates (Gauss)) *Given any H^s metric g on M, there exists a local coordinate chart $\varphi : U \to \mathbb{R}^2$ about any $x \in M$ such that for $\psi = \varphi^{-1}$, $\psi^* g = \lambda \cdot g_{eucl}$, where λ is a positive H^s function and g_{eucl} is the euclidean metric on $\mathbb{R}^2$.*

We shall need some basic results on group actions on manifolds.

Let $\mathcal{G}$ be a C^∞ (Hilbert-)Lie group acting by right action on a C^∞ (Hilbert) manifold $\mathcal{N}$: $A : \mathcal{G} \times \mathcal{N} \to \mathcal{N}, (a, x) \mapsto x \cdot a$ the orbit of $\mathcal{G}$ through x is defined to be $\mathcal{G}_x := \{x \cdot a \mid a \in \mathcal{G}\}$.

Definition 1.3.1   1) The action is proper, iff the map

$$\tilde{A} : \mathcal{G} \times \mathcal{N} \to \mathcal{N} \times \mathcal{N}, (a, x) \mapsto (x \cdot a, x)$$

is proper (i.e. if pre-images of compact sets are compact).

2) The action is free, iff it has no fixed points, i.e.

$$(\exists x : x \cdot a = x) \Rightarrow a = id \ .$$

Theorem 1.3.2 *Let a C^∞ Hilbert Lie group $\mathcal{G}$ act (by right action) on a C^∞ Hilbert manifold $\mathcal{N}$. If the action is smooth, proper, and free, then:*

(i) For all $x \in \mathcal{N}$, $\mathcal{G}_x$ is a smooth, closed submanifold of $\mathcal{N}$

(ii) The quotient space $\mathcal{N}/\mathcal{G}$ is a smooth manifold

(iii) Its tangent space $T_{[x]}(\mathcal{N}/\mathcal{G})$ can be identified with any subspace of $T_x\mathcal{N}$ complementary to $T_x\mathcal{G}_x$

(iv) The quotient map $\pi : \mathcal{N} \rightarrow \mathcal{N}/\mathcal{G}$ is a C^∞ submersion (this means: $D\pi(x) : T_x\mathcal{N} \rightarrow T_{[x]}\mathcal{N}/\mathcal{G}$ is surjective).

In a way this theorem contains the basic idea of Teichmüller theory. One wants to divide out a group $\mathcal{D}_0$ from a manifold $\mathcal{A}$ to get a quotient manifold $\mathcal{T}$. However, technical reasons prevent us from applying the theorem to this situation: $\mathcal{D}_0$ is not a Hilbert Lie group. As a manifold, it is modelled after the linear space of all C^∞ vector fields on M, which is not even a Banach space. In this situation, a corresponding theorem is not available.

On the other hand, $\mathcal{D}_0^{s+1}$ is a Hilbert Lie group, and it does act on the Hilbert manifold $\mathcal{A}^s$. Indeed, if $f \in \mathcal{D}_0^{s+1}, J \in \mathcal{A}^s$, then $f^*J \in \mathcal{A}^s$, since

$$(f^*J)_x = df_x^{-1} J_{f(x)} df_x \ ,$$

but the action is not smooth: trying to differentiate the above formula f forces the derivative of J_y with respect to y to enter. It is in $\mathcal{H}^{s-1}(T_2^1 M)$, but no longer of differentiability class $\mathcal{H}^s$. So we have a theorem available, but it does not apply to the situation at hand.

Nevertheless, the theorem does apply in a different situation, which will be useful for our purpose:

Let $\mathcal{P}^s$ be the set of all $\mathcal{H}^s$ positive functions on M. It is an open subset of the Hilbert space $H^s(M)$ (provided $s > 1$), and thus a Hilbert manifold. Ordinary multiplication of functions gives it the structure of a C^∞ Hilbert Lie group. It acts on $\mathcal{M}^s$ by $(p,g) \mapsto p \cdot g$. This action is obviously smooth and free. One also immediately checks that it is proper.

This gives us the following

Corollary 1.3.3 *The quotient $\mathcal{M}^{\bullet}/\mathcal{P}^{\bullet}$ has the structure of a C^{∞} Hilbert manifold, and its tangent space can be identified with $(S_2^{\bullet}(g))^T$.*

PROOF: The splitting of lemma 1.2.1 identifies $(S_2^{\bullet}(g))^T$ with the orthogonal complement of the tangent space of the orbit of $\mathcal{P}^{\bullet}$. ∎

$\mathcal{D}$ and $\mathcal{D}_0$ act on $\mathcal{M}^{\bullet}/\mathcal{P}^{\bullet}$ and on $\mathcal{M}/\mathcal{P}$ by $f^*[g] := [f^*g]$, which is well-defined, as one easily sees. The point is now the following

Theorem 1.3.4 *There exists a smooth diffeomorphism of Hilbert manifolds:*
$\Phi : \mathcal{M}^{\bullet}/\mathcal{P}^{\bullet} \to \mathcal{A}^{\bullet}$. *$\Phi$ takes $\mathcal{M}/\mathcal{P}$ onto $\mathcal{A}$.*

PROOF: Φ is defined first as a mapping from $\mathcal{M}^{\bullet}$ to $\mathcal{A}^{\bullet}$. It will follow soon afterwards that one can pass to the quotient. Define $\Phi : \mathcal{M}^{\bullet} \to \mathcal{A}^{\bullet}, g \mapsto -g^{-1}\mu_g$, in other words

$$g(x)(u, \Phi(g)v) = -\mu_g(x)(u, v)$$
$$\Phi(g)^i{}_j = -g^{ik}(\mu_g)_{kj} .$$

The geometrical meaning of Φ is very clear: once a metric g is given, angles are a well-defined concept. Moreover M was assumed to be oriented. Since M is 2-dimensional, this is enough information to define the mapping $\Phi(g)$ of the tangential spaces T_xM into themselves to be counterclockwise rotation by a right angle. The map Φ has been known for a long time as a map between the *sets* $\mathcal{M}^{\bullet}$ and $\mathcal{A}^{\bullet}$.

However, it is the key of this approach to Teichmüller theory to consider it as a map between the *manifolds* $\mathcal{M}^{\bullet}$ and $\mathcal{A}^{\bullet}$, i.e. to consider its derivative as well.

Φ has the following properties:

(i) Φ is well-defined as a map from $\mathcal{M}^{\bullet}$ to $\mathcal{A}^{\bullet}$ (and from $\mathcal{M}$ to $\mathcal{A}$), i.e. $\Phi(g) \in \mathcal{A}^{\bullet}$ if $g \in \mathcal{M}^{\bullet}$

(ii) $J = \Phi(g)$ is Hermitian with respect to g, i.e. $g(Ju, Jv) = g(u, v)$

(iii) $\Phi(p \cdot g) = \Phi(g) \quad (\forall p \in \mathcal{P}^{\bullet})$

(iv) If $\Phi(g_1) = \Phi(g_2)$, then $g_1 = p \cdot g_2$ for some $p \in \mathcal{P}^{\bullet}$

(v) Φ is onto $\mathcal{A}^{\bullet}$

(vi) $\Phi : \mathcal{M}^s \to \mathcal{A}^s$ is a C^∞ submersion, i.e. $\Phi \in C^\infty$ and $D\Phi(g) : T_g\mathcal{M}^s \to T_{\Phi(g)}\mathcal{A}^s$ is surjective. Moreover,

$$\ker D\Phi(g) = (S_2^s(g))^c .$$

Putting these properties together proves the theorem by inspection.

[Note that inspection is only the second most powerful method of proof, the most powerful being intimidation!]

(i) becomes a trivial calculation in local conformal coordinates:

$$g_{ij} = \lambda \cdot \delta_{ij} , \quad g^{ik} = \frac{1}{\lambda}\delta^{ik} ,$$

$$((\mu_{ij})) = \begin{bmatrix} 0 & \lambda \\ -\lambda & 0 \end{bmatrix} , \quad ((\Phi(g)^i{}_j)) = \begin{bmatrix} 0 & -1 \\ 1 & 0 \end{bmatrix} .$$

The same calculation proves (iii) as well.

(ii) is also obvious in local conformal coordinates.

(iv) Since the space of antisymmetric 2×2 matrices is one dimensional, there exists a function p such that $\mu_{g_1} = p \cdot \mu_{g_2}$. p is clearly in $H^s(M)$. Then $g_1^{-1}\mu_{g_1} = g_2^{-1}\mu_{g_2}$ implies $g_1 = p\,g_2$.

(v) Given some J, we must produce a g such that $\Phi(g) = J$. To this end start with an arbitrary $\hat{g} \in \mathcal{M}^s$ and define a new metric g by

$$g(x)(u,v) := \hat{g}(x)(u,v) + \hat{g}(x)(J_x u, J_x v) .$$

Then $g(JX, JX) = g(X,X)$ and $g(X, JX) = 0$. Therefore

$$\mu_g(X, JX) = \sqrt{\det \begin{bmatrix} g(X,X) & g(X,JX) \\ g(JX,X) & g(JX,JX) \end{bmatrix}} = g(X,X) .$$

Since $\{X_x, (JX)_x\}$ forms a basis of T_xM, any vector field Y can be written as $Y = a\,X + b\,JX$ with functions a, b. Then, $\mu_g(X,Y) = a\mu_g(X,X) + b\mu_g(X,JX) = b\mu_g(X,JX)$.

$$g(X, JY) = a\,g(X,JX) - b\,g(X,X) = -b\,g(X,X) = -b\mu_g(X,JX) .$$

So, $g(X, JY) = -\mu_g(X,Y)$, which is just the defining property for $\Phi(g) = J$.

(vi) Let $h \in T_g\mathcal{M}$, its components in local coordinates being h_{ij}. Differentiation of $g_{ij}g^{jk} = \delta_i^k$ shows that the map $\xi^{ij} : g \mapsto g^{ij}$ has the derivative $D\xi^{ij}(g) : h \mapsto -h^{ij}$.

The derivative of $\mu(g) = \pm\sqrt{\det((g_{ij}))}$ is

$$
\begin{aligned}
D\mu(g) \quad : \quad h \mapsto & \frac{1}{\pm 2\sqrt{\det((g_{ij}))}} \left(\det\begin{bmatrix} h_{11} & g_{21} \\ h_{21} & g_{22} \end{bmatrix} + \det\begin{bmatrix} g_{11} & h_{21} \\ g_{21} & h_{22} \end{bmatrix} \right) \\
= & \pm\frac{\sqrt{\det((g_{ij}))}}{2} \left(\det\left(\begin{bmatrix} g^{11} & g^{12} \\ g^{21} & g^{22} \end{bmatrix} \begin{bmatrix} h_{11} & g_{21} \\ h_{21} & g_{22} \end{bmatrix} \right) + \cdots \right) \\
= & \tfrac{1}{2}\mu(g) \left(\det\begin{bmatrix} h_1^1 & 0 \\ h_1^2 & 1 \end{bmatrix} + \det\begin{bmatrix} 1 & h_2^1 \\ 0 & h_2^2 \end{bmatrix} \right) \\
= & \tfrac{1}{2}\mu(g)tr_g h \ .
\end{aligned}
$$

Therefore, the derivative of $\Phi(g)^i{}_j = -g^{ik}\mu(g)_{kj} =: J^i{}_j$ is

$$
\begin{aligned}
D\Phi(g)^i{}_j : h \mapsto h^{ik}\mu(g)_{kj} - g^{ik}\tfrac{1}{2}\mu(g)_{kj}tr_g h &= h^i{}_k\mu(g)^k{}_j + \tfrac{1}{2}J^i{}_j tr_g h \\
&= -h^i{}_k J^k{}_j + \tfrac{1}{2}J^i{}_j tr_g h \\
&= -\left[\left(H - \tfrac{1}{2}(tr\ H)I \right) J \right]^i{}_j
\end{aligned}
$$

with $H = ((h^i{}_j))$.

Thus, since J is invertible,

$$
\ker D\Phi(g) = \left\{ h \mid H = \frac{1}{2}tr\ H \cdot I \right\} = S_2^{\bullet}(g)^c
$$

as claimed.

To show that $D\Phi(g)$ is surjective means to show that $\left(H - \tfrac{1}{2}(tr\ H) \cdot I \right) J$ runs through all of $T_J\mathcal{A} \cong \{K \mid KJ + JK = 0\}$ as H runs trough all of $\{H \mid tr\ H = 0\}$.

(Remember that $T_g\mathcal{M}^{\bullet} = S_2^{\bullet}(g)^c \oplus S_2^{\bullet}(g)^T$.)

Since for $tr\ H = 0$, $D\Phi(g)h = -HJ$ and $H \mapsto -HJ$ is an isomorphism of $T_J\mathcal{A}^{\bullet}$ to itself, the surjectivity immediately follows from

Lemma 1.3.5 *Given* $J \in \mathcal{A}^{\bullet}$, *then*

$$
HJ = -JH \qquad \Leftrightarrow \qquad tr\ H = 0 \ and \ H \ is \ g\text{-}symmetric \ .
$$

This ends the proof of the theorem since $H = g^{-1}h$ must by definition be g-symmetric and hence in $T_J\mathcal{A}^{\bullet}$. ∎

PROOF OF THE LEMMA: Work in conformal coordinates, so $J = \begin{bmatrix} 0 & -1 \\ 1 & 0 \end{bmatrix}$. If $tr\, H = 0$ and H is symmetric, then $H = \begin{bmatrix} a & b \\ b & -a \end{bmatrix}$. $HJ + JH = 0$ follows. If $HJ = -JH$, then $tr\, H = -tr\, J^2 H = -tr\, JHJ = -tr\, J(-JH) = tr\, J^2 H = -tr\, H$.

Thus $tr\, H = 0$, and again a calculation in coordinates shows that H is symmetric. ∎

To sum up, we know that $\mathcal{D}^{s+1}$ acts on $\mathcal{M}^s, \mathcal{A}^s, \mathcal{M}^s/\mathcal{P}^s$, and that $\Phi : \mathcal{M}^s/\mathcal{P}^s \to \mathcal{A}^s$ is a diffeomorphism. Moreover, our proof shows in fact that Φ is a bijective map between $\mathcal{M}/\mathcal{P}$ and $\mathcal{A}$.

Finally, we have

Lemma 1.3.6 $\Phi : \mathcal{M}^s/\mathcal{P}^s \to \mathcal{A}^s$ *is* $\mathcal{D}^{s+1}$-*equivariant.*

PROOF:

$$\Phi\left(f^*[g]\right) = \Phi\left([f^*g]\right) = -(f^*g)^{-1}\mu_{f^*g} = -(f^*g)^{-1}f^*\mu_g = -f^*(g^{-1}\mu_g) = f^*\Phi(g)$$

∎

We can now complete the proof of theorem 1.1.1 as well: We had left out the surjectivity of the map $\Gamma : \mathcal{C}^s \to \mathcal{A}^s$. Now, given any almost complex structure $J \in \mathcal{A}^s$, a class $[g]$ of metrics corresponds to it: $[g] = \Phi^{-1}(J)$. According to theorem 1.3.1 we can choose conformal coordinates for any $\tilde{g} \in [g]$; conformality of coordinates obviously does not depend on the chosen representative $\tilde{g}$, but only on its conformal class $[\tilde{g}] = [g]$. Given two conformal coordinate charts, $\varphi : U_\varphi \to \mathbb{R}^2, \psi : U_\psi \to \mathbb{R}^2$, $\varphi \circ \psi^{-1}$ is a conformal (and orientation preserving) mapping, where defined. So it is holomorphic, and therefore the set of all conformal coordinate charts gives the complex structure c. In conformal coordinates (x,y), we have $g = \rho \cdot (dx^2 + dy^2)$, and the corresponding matrices are $\begin{bmatrix} \rho & 0 \\ 0 & \rho \end{bmatrix}$ for g, $\begin{bmatrix} 0 & \rho \\ -\rho & 0 \end{bmatrix}$ for μ_g, thus $\begin{bmatrix} 0 & -1 \\ 1 & 0 \end{bmatrix}$ for J. Using equation (1.1) and noting, that $d\varphi_x = \begin{bmatrix} \lambda & 0 \\ 0 & \lambda \end{bmatrix}$ in conformal coordinates (for some λ), it is clear that $\Gamma c = J$. ∎

So we have an equivalence between $\mathcal{A}^s/\mathcal{D}^{s+1}$ and $(\mathcal{M}^s/\mathcal{P}^s)/\mathcal{D}^{s+1}$. Our next goal is to replace the quotient $\mathcal{M}^s/\mathcal{P}^s$ by something simpler, namely a C^∞ submanifold of $\mathcal{M}^s$. This is done choosing a unique representative in $\mathcal{M}^s/\mathcal{P}^s$ in a natural way: it is the one metric of constant curvature -1 in the whole conformal class. Let us repeat the necessary facts from differential geometry:

Given a metric $g \in \mathcal{M}^{s+2}$ on a manifold M, there is a function $R(g) \in H^s$ associated to it, called the scalar curvature. R is twice the Gauss curvature in case $\dim M = 2$. The reader can find the formula in appendix A. It will not be needed here.

In the case of an embedded (orientable) surface $M \subset \mathbb{R}^3$, where M inherits its metric from $\mathbb{R}^3$ by $g(x)(X_x, Y_x) := \langle X_x, Y_x \rangle_{\mathbb{R}^3}$, the Gauss curvature $K(g)$ has the following geometric meaning:

Let $n : M \to S^2 \hookrightarrow \mathbb{R}^3$ be the normal map, i.e. $n(x)$ is the outward normal unit vector at x,

$$dn(x) : T_x M \to \mathbb{R}^3 \ ,$$

but actually $dn(x) \perp n(x)$, so $dn(x)$ maps into $T_x M$. In this case $R(g)(x) = 2 \det dn(x)$.

The complicated formula in appendix A has the advantage of giving $R(g)$ only in terms of g, thus independently of any embedding.

We will need the following two theorems:

Theorem 1.3.7 *(Poincaré)* *Suppose M is a compact oriented surface of genus > 1. Then given any $g \in \mathcal{M}^s$, there exists a unique $\lambda \in \mathcal{P}^s$ such that $R(\lambda g) \equiv -1$.*

Theorem 1.3.8 *(Gauss-Bonnet)* *M is a compact oriented surface. Then*

$$\int\limits_M R(g) d\mu_g = 4\pi \chi(M) = 4\pi(2 - 2 \text{ genus } (M)) \ .$$

Both theorems are well known in differential geometry. A proof for the first one, which is different from Poincaré's proof, will be given soon.

Let $\mathcal{M}^s_{-1} := \{g \in \mathcal{M}^s \mid R(g) \equiv -1\}$.

Poincaré's theorem immediately gives a bijective correspondence between $\mathcal{M}^s/\mathcal{P}^s$ and $\mathcal{M}^s_{-1}$. We will show later that this correspondence is a diffeomorphism, and indeed $\mathcal{D}^{s+1}$-equivariant (for an action of $\mathcal{D}^{s+1}$ on $\mathcal{M}^s_{-1}$ yet to be constructed).

But before continuing we have to consider some differential operators which will play a role in the sequel.

1.4   Some Differential Operators and their Adjoints

Let $v : M \to \mathbb{R}$ be a differentiable function. The gradient of v is defined with respect to a metric on M (and cannot be defined unless M is equipped with a metric):
$\nabla_g v : M \to TM, \ (\nabla_g(x))^i = g^{ij}(x)\frac{\partial v}{\partial x^j}$.

∇_g is a differential operator which maps functions to vector fields.

Since we generally assume M to be equipped with a metric g, there are L^2 inner products defined in a natural way on spaces of functions, vector fields, and tensors, respectively:

For functions v, w on M,
$$\langle v, w \rangle_{L^2} := \int_M v \cdot w \, d\mu_g \ .$$
For vector fields X, Y on M,
$$\langle X, Y \rangle_{L^2} := \int_M g(x)(X_x, Y_x) \, d\mu_g \ .$$
For 0-2 tensors h, k, let H, K be the corresponding 1-1 tensors
$$H_l^i = g^{ij} H_{jl}, \ \ K_l^i = g^{ij} k_{jl} \ .$$

It is with respect to this scalar product that, given a differential operator, its adjoint differential operator is defined. We will have to use the divergence operator δ_g, which is minus the adjoint of the gradient.
$$\delta_g : \mathcal{X}^s(M) \to H^{s-1}(M) \ ,$$
where $\mathcal{X}^s(M)$ represents the H^s vector fields on M and $H^{s-1}(M)$ are the H^{s-1} smooth functions. δ_g is given in coordinates by $\delta_g X = \frac{1}{\sqrt{g}}\frac{\partial}{\partial x^i}\left(X^i \sqrt{g}\right)$.

It is merely integration by parts which shows that this is indeed the negative adjoint of the gradient, i.e. that
$$\langle \delta_g X, v \rangle_{L^2} = -\langle X, \nabla_g v \rangle_{L^2} \ .$$
It is a standard matter to use the metric in order to define the divergence of forms and tensors:

If ω is a 1-form, then let X_ω be its associated vector field and define $\delta_g \omega := \delta_g X_\omega$.

The divergence of a g-symmetric 1-1 tensor H is the 1-form
$$\frac{1}{\sqrt{g}}\frac{\partial}{\partial x^j}\left(h_i^j \sqrt{g}\right) - \frac{1}{2}g^{kl}h_l^j\frac{\partial g_{jk}}{\partial x^i} \ .$$

The divergence of a symmetric 0-2 tensor is the divergence of its associated 1-1 tensor. By raising an index we can also think of the divergence of a 1-1 tensor or a 0-2 tensor as a vector field, and we use *both conventions* in the text.

One can now combine the divergence and the gradient operators in the same way as one does in $\mathbb{R}^3$. There, the standard Laplacian is defined by $\Delta v = \text{div } \nabla v$. The natural generalization on a Riemannian manifold is

$$\Delta_g v := \delta_g \nabla_g v = \delta_g \delta_g^* v \ .$$

This is called the Laplace-Beltrami operator. It is a second order differential operator and, being the product of an operator and its adjoint, it is self-adjoint. (Let it be enough to say it is formally self-adjoint; we have no need here to go into the theory of self-adjoint operators on Hilbert spaces.)

In local coordinates $\Delta_g v = \frac{1}{\sqrt{g}} \frac{\partial}{\partial x^i} \left(g^{ij} \sqrt{g} \frac{\partial v}{\partial x^j} \right)$.

Proposition 1.4.1 *Given any H^s-vector field X, it can be written as $X = X^0 + \nabla_g v$, where X^0 is a divergence free ($\delta_g X^0 = 0$) H^s vector field, $v \in H^{s+1}(M, \mathbb{R})$.*

PROOF: If a decomposition as claimed exists, it follows $\delta_g X = \Delta_g v$. In order to produce v, one has to solve this equation for v. One must therefore show that $\delta_g X$ is in the range of Δ_g. Now, since Δ_g is a self-adjoint operator, its range is the orthogonal complement of its kernel. So, we have to show $\delta_g X \perp \ker \Delta_g$.

If $w \in \ker \Delta_g$, then $0 = \int \Delta_g w \cdot w = -\int g(x)(\nabla_g w, \nabla_g w) d\mu_g$, so w is constant. But then $\langle \delta_g X, w \rangle = -\langle X, \nabla_g w \rangle = 0$. ∎

Now let $f_t \in \mathcal{D}_0$, $-\varepsilon < t < \varepsilon$, $f_0 = id$ be a differentiable family of diffeomorphisms. Then $\frac{df_t}{dt}\big|_{t=0} =: X$ is a vector field.

The Lie derivative $L_X g$ of a metric $g \in \mathcal{M}^s$ is defined by

$$L_X g := \frac{d}{dt}\Big|_{t=0} f_t^* g \ . \tag{1.3}$$

It can be expressed in local coordinates as follows:

$$(L_X g)_{ij} = X^k \frac{\partial g_{ij}}{\partial x^k} + g_{ik} \frac{\partial X^k}{\partial x^j} + g_{jk} \frac{\partial X^k}{\partial x^i} \ .$$

The second formula shows that $L_X g$ does not depend on f_t but only on X. The first one shows that it does not depend on a choice of coordinates. More information can be found in books on differential geometry, e.g. Spivak [103]. The reason for introducing the Lie derivative is that we want to prove that $\mathcal{T}(M) \cong \mathcal{M}_{-1}/\mathcal{D}_0$, and is a C^∞ manifold. It will therefore be important for us to note that the tangent vectors to the orbit of $\mathcal{D}_0$ are vectors of the form $L_X g$.

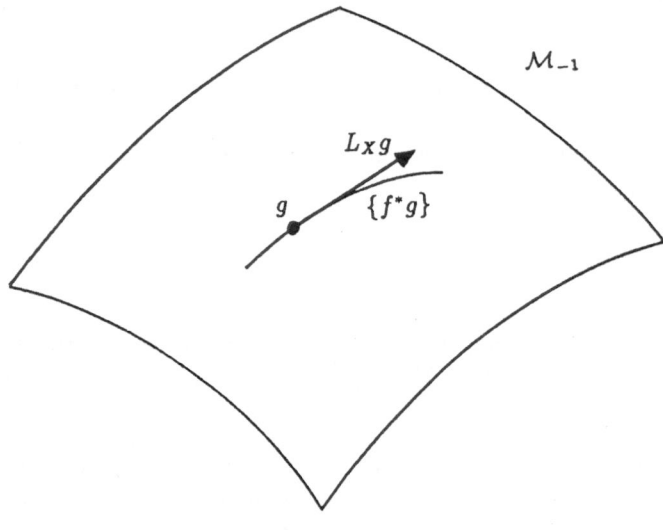

Figure 1.1

Given the metric g, the Lie derivative gives us a mapping

$$\alpha_g : X \mapsto \alpha_g(X) := L_X g, \quad \mathcal{X}(M) \to S_2(M) \ ,$$

where $\mathcal{X}(M)$ denotes the space of vector fields on M. It is the L^2-adjoint of $-\delta_g$:

$$\langle \alpha_g(X), h \rangle_{L^2} = -\langle X, \delta_g h \rangle$$

($\delta_g h$ is to be understood as a vector field rather than a form by raising an index).

Now, $\delta_g \delta_g^*$ is a second order self-adjoint elliptic operator defined on 0-2 tensors. We have a similar theorem for 0-2 tensors as the one above for vectors.

Theorem 1.4.2 *Assume, $g \in \mathcal{M}^{s+1}$. Then, given any 0-2 tensor $h \in S_2^s$ there exists an L^2-orthogonal splitting $h = h^0 + L_X g$ where $\delta_g h^0 = 0$ and h and $L_X g$ are in S_2^s.*

Remarks 1.4.1 *Later (theorem 2.2.2), we will show that X is uniquely determined in the case of genus$(M) > 1$. The assumption that $g \in \mathcal{M}^{s+1}$ is technical and is needed to assure that elliptic theory applies to the operator $\delta_g \delta_g^*$.*

PROOF: Given h, we try to solve $\delta_g \delta_g^* X = -\delta_g h$. This is possible if and only if $\delta_g h \perp \ker \delta_g \delta_g^*$.

Now $\delta_g \delta_g^* X = 0$ implies $L_X g = 0$ (take the scalar product with X and integrate by parts).

Therefore, for any $X \in \ker \delta_g \delta_g^*$, we have

$$\langle \delta_g h, X \rangle = -\langle h, L_X g \rangle = 0 \ .$$

Now, setting $h^0 := h - L_X g$, it immediately follows $\delta_g h^0 = 0$. ∎

1.5 Proof of Poincaré's Theorem

To appreciate the spirit of this proof as well as to understand some of its technical details needs some acquaintance with the direct methods of calculus of variations. But also without such knowledge, the reader can follow large parts of the proof. On the other hand, the proof can be skipped with no harm, if one is willing to accept the theorem.

First note, that the horrible formula in the appendix for $R(g)$ becomes quite nice, if local conformal coordinates are chosen on the surface M, i.e. $g_{ij} = \lambda \cdot \delta_{ij}$.

In these coordinates, we have

$$R(g) = -\frac{1}{\lambda} \Delta \log \lambda \quad \left(\Delta = \frac{\partial^2}{\partial x^2} + \frac{\partial^2}{\partial y^2} \right)$$

and $\Delta_g = \frac{1}{\lambda} \Delta$.

We have to find ρ such that $R(\rho \cdot g) \equiv -1$; since $\rho > 0$ we may write $\rho = e^v$. Now,

$$R(e^v g) = -\frac{1}{\lambda e^v} \Delta \log(e^v \lambda) = \frac{1}{e^v} \left(-\frac{1}{\lambda} \Delta \log \lambda - \frac{1}{\lambda} \Delta v \right) \ .$$

So now abandoning conformal coordinates we have to solve the global intrinsically defined PDE

$$-\Delta_g v + R(g) + e^v = 0 \ . \tag{1.4}$$

This can be achieved by minimizing the following functional:

$$I(v) := \frac{1}{2} \int_M g(x)(\nabla_g v, \nabla_g v) d\mu_g + \int (R(g)v + e^v) \, d\mu_g$$

The first of the terms on the right is called Dirichlet's energy, $E(g, v)$.

Formally, (1.4) is the Euler-Lagrange equation of the functional I. If one can prove existence of a minimum of I in some space (and the differentiability of I), then the derivative of I at that point is 0, hence the equation (1.4) holds. Indeed,

$$\begin{aligned} DI(v)h &= \int g(x)(\nabla_g v, \nabla_g h) d\mu_g + \int (R(g) + e^v) h \, d\mu_g \\ &= \int (-\Delta_g v + R(g) + e^v) h \, d\mu_g \ . \end{aligned}$$

If this is 0 for all h, the term in parentheses has to vanish. But it is difficult to minimize I on $H^1(M)$ directly. $H^1(M)$ *is* the right space to work in, but the difficulty stems from the fact that e^v need not be integrable for general $v \in H^1(M)$, so I is not finite on all of $H^1(M)$.

The device to help us out would be called an a-priori estimate in partial differential equations. For the purpose of motivation, assume that x_0 is a point where a supposed differentiable minimizer v has a maximum. (Sure, we haven't v yet, let alone its smoothness, but never mind, this is heuristic only.)

At such a point, $-\Delta_g v(x_0) \geq 0$, so (1.4) implies that

$$e^{v(x_0)} \leq -R(g)(x_0) \leq -\min R(g) \leq |\min R(g)| \ .$$

The rightmost term is not 0, since at some point, $R(g)$ must be negative due to Gauss-Bonnet (indeed at x_0 by the first inequality). So $v(x) \leq v(x_0) \leq \log |\min R(g)|) =: \xi$ is a bound for v. It depends only on (M, g).

Let $C := \{v \in H^1(M, \mathbb{R}) \mid E(g, v) < \infty, \ v(x) \leq 1 + \xi \ \text{a.e.}\}$.

The goal is to show that I has a minimum v on C, and that v actually satisfies the Euler-Lagrange equation.

To this end, we show first that I is bounded below on C and that every minimizing sequence v_n (i.e. $I(v_n) \to \inf I(v)$) is bounded in a norm which is equivalent to the H^1 norm.

Dirichlet's energy alone cannot be used as a norm, because it is 0 for constant functions. This corresponds to the fact that the first eigenvalue of $-\Delta_g$ on M is 0. The second eigenvalue, λ_2, is positive, and it holds:

$$E(g,u) \geq \frac{\lambda_2}{2} \int u^2 d\mu_g \quad (\forall u \text{ such that } \int_M u \, d\mu_g = 0) \ .$$

So, we write $u = u_0 + m$, with $\int u_0 d\mu_g = 0$ and m a constant, namely the mean value of u on M:

$$m = \frac{\int\limits_M u \, d\mu_g}{\int\limits_M d\mu_g}$$

The norm to be used is

$$||u|| := \sqrt{E(g,u)} + |m| \ .$$

We now have the estimate

$$
\begin{aligned}
I(u) &= E(g,u_0) + \int R(g)u_0 d\mu_g + \int R(g)m d\mu_g \\
&= E(g,u_0) + \int R(g)u_0 d\mu_g + 4\pi m \chi(M) \\
&\geq E(g,u_0) - \int \left(\tfrac{\lambda_2}{4}u_0^2 + \tfrac{1}{\lambda_2}R(g)^2\right) d\mu_g + 4\pi m \chi(M) \\
&\geq \tfrac{1}{2}E(g,u_0) - \tfrac{1}{\lambda_2} \int R(g)^2 d\mu_g + 4\pi m \chi(M) \ .
\end{aligned}
$$

The second term is merely a constant depending only on (M,g), but the last term needs estimating. Note that $\chi(M) < 0$.

__Case 1:__ If $m > 0$, then $0 < |m| = m < 1 + \xi$. So

$$I(u) \geq \frac{1}{2}E(g,u_0) - \frac{1}{\lambda_2} \int_M R(g)^2 d\mu_g - 4\pi(1+\xi)\,|\chi(M)| \ ,$$

which gives a lower bound for I (since $E(g,u_0) \geq 0$) and an upper bound for $E(g,u_0)$ in case u runs through a minimizing sequence (we have just had the bound for $|m|$).

__Case 2:__ If $m \leq 0$, then
$I(u) \geq \tfrac{1}{2}E(g,u_0) + 4\pi|m|\,|\chi(M)| - \tfrac{1}{\lambda_2}\int R(g)^2 d\mu_g$, which again gives a lower bound for I, and an upper bound for $||u||$ if u runs through a minimizing sequence.

Now, standard arguments guarantee the existence of a minimizer v on C. (Given a minimizing sequence, select a weakly and pointwise a.e. convergent subsequence; $E(g,\cdot)$ is weakly lower semicontinuous; by Fatou's lemma, e^v is lower semicontinuous with respect to pointwise a.e. convergence.)

We now have to show that the (any) minimizer v is in the interior of C (for there is no chance to get the Euler-Lagrange equation for minimizers at the boundary). To this end, the former heuristic result $v \leq \xi$ is now shown to hold.

Let $w \in C$ be arbitrary. C is convex. So the minimizing property gives
$$I(v) \leq I\left(tw + (1-t)v\right) = I\left(v + t(w-v)\right) \quad \forall t \in [0,1] \ .$$

Therefore $\frac{d}{dt}\Big|_{t=0} I\left(v + t(w-v)\right) \geq 0$.

(One can easily see that this derivative does exist.)

So,
$$\int_M g(x)\left(\nabla_g v, \nabla_g(v-w)\right) d\mu_g + \int_M (R(g) + e^v) \cdot (v-w) d\mu_g \leq 0 \ .$$

This holds for any $w \in C$; take $w := \min(v, \xi)$ as an appropriate choice.

Then both terms under the integral signs vanish at points x where $v(x) \leq \xi$. At other points, w is constant, so $\nabla_g w = 0$ there. Thus
$$\int_{\{v(x) > \xi\}} g(x)(\nabla_g v, \nabla_g v) d\mu_g + \int_{\{v(x) > \xi\}} (R(g) + e^v) \cdot (v - \xi) d\mu_g \leq 0 \ .$$

By definition of ξ, $R(g) + e^\xi \geq 0$, so $R(g) + e^v > 0$ on the domain of integration. All terms under the integral signs are > 0, so the inequality cannot be satisfied unless the domain of integration has measure 0. This proves that $v(x) \leq \xi$ a.e.

Still some work is needed to show that v has higher differentiability than H^1 (so that partial integration which was used to get (1.4) is justified) and that v satisfies (1.4) in the classical sense. However, this is standard, given the estimates above.

The detailed argument ("H^s estimates for the Laplacian") can be found, say, in Gilbarg-Trudinger [42], theorems 8.9-8.11.

The key idea is that
$$\Delta_g v = R(g) + e^v$$
holds in the weak sense.

For $g \in \mathcal{H}^s, R(g) \in H^{s-2}$, and e^v is known to be in L^2. The right hand side being in L^2, v is in H^2. Therefore so is the right hand side, implying $v \in H^4$ (if $s \geq 4$). The induction argument ends with $v \in H^s$ when it is $R(g)$ that determines the smoothness of the right hand side.

Now the maximum principle can be used to show that (1.4) has at most one solution, so it is unique. Indeed, let $v_1 - v_2$ be solutions in $H^s \subset C^0$ ($s \geq 3$). Then subtracting yields $\Delta_g(v_1 - v_2) = e^{v_1} - e^{v_2}$.

Let x_0 be a point where $v_1(x) - v_2(x)$ attains its maximum on M. (M is compact, and $v_1 - v_2$ is continuous, so x_0 exists.)

At a maximum point, $\Delta_g(v_1 - v_2)(x_0) \le 0$, so $e^{v_1(x_0)} \le e^{v_2(x_0)}$. Therefore $v_1 - v_2$ is ≤ 0 at its maximum point, thus everywhere. By symmetry, $v_1 - v_2 \ge 0$, too. This completes the proof.   ∎

1.6 The Manifold $\mathcal{M}^s_{-1}$ and the Diffeomorphism with $\mathcal{M}^s/\mathcal{P}^s$

We have already seen that $\mathcal{M}^s/\mathcal{P}^s \cong \mathcal{A}^s$ by a $\mathcal{D}$-equivariant diffeomorphism. Poincaré's theorem gives us a bijective map between the sets $\mathcal{M}^s/\mathcal{P}^s$ and $\mathcal{M}^s_{-1}$. The next step is naturally

Theorem 1.6.1 $\mathcal{M}^s_{-1}$ *is a smooth submanifold of* $\mathcal{M}^s$.

PROOF: The scalar curvature is a map

$$R : \mathcal{M}^s \rightarrow H^{s-2}(M, \mathbb{R}) \ ,$$

and $\mathcal{M}^s_{-1} = R^{-1}(-1)$. By the implicit function theorem, it suffices to show that -1 is a regular value for R, i.e. that if $R(g_0) = -1$ for some g_0, then

$$DR(g_0) : T_g \mathcal{M}^s \approx S^s_2 \rightarrow H^{s-2}(M, R)$$

is surjective. Luckily, there is an easy formula for $DR(g_0)$ in 2 dimensions at points where $R(g_0) = -1$. It is:

$$DR(g_0)h = -\Delta_{g_0}(tr_{g_0}h) + \delta_{g_0}\delta_{g_0}h + \frac{1}{2}tr_{g_0}h \ . \qquad (1.5)$$

Since this formula is so central to our presentation, we present a proof in the appendix.

Once we have established the claim of surjectivity of $DR(g_0)$, the implicit function theorem says that $T_{g_0}\mathcal{M}^s_{-1} = \ker DR(g_0)$. This gives us a hint for the proof of surjectivity, namely that we should consider those h which are in some sense transversal to $\mathcal{M}_{-1}$ (say pointing in direction of the orbit of $\mathcal{P}^s$). Therefore, we consider only h of the form $h = \lambda \cdot g_0$ ($\lambda \in H^s$). It is sufficient to show surjectivity when h runs through this restricted set. An easy calculation shows that $\delta_{g_0}\delta_{g_0}(\lambda_{g_0}) = \Delta_{g_0}\lambda$. Inserting this yields

$$DR(g_0)\lambda g_0 = -\Delta_{g_0}\lambda + \lambda \ .$$

This map is well known in elliptic theory to be surjective from $H^s(M, \mathbb{R})$ to $H^{s-2}(M, \mathbb{R})$. Let us repeat the core of the proof for this.

The Fredholm alternative which holds for $-\Delta + id$ in elliptic theory says that surjectivity follows from injectivity. But if $-\Delta\lambda + \lambda = 0$, multiplying by λ and integrating by parts yields

$$\int_M g(x)(\nabla_g\lambda, \nabla_g\lambda)d\mu_g + \int_M \lambda^2 d\mu_g = 0 \ ,$$

showing $\lambda = 0$. ∎

This ends the proof and has almost proved

Theorem 1.6.2 $\mathcal{M}^s_{-1}$ *is diffeomorphic to* $\mathcal{M}^s/\mathcal{P}^s$.

PROOF: Let $\pi : \mathcal{M}^s \to \mathcal{M}^s/\mathcal{P}^s$ be the quotient map.

$\pi_{-1} := \pi|\mathcal{M}^s_{-1}$ is surjective by the existence part of Poincaré's theorem and injective by the uniqueness part.

To show that π_{-1} is a diffeomorphism, we have to show that $D\pi_{-1}$ is an isomorphism.

$$\begin{aligned}
T_{[g]}\mathcal{M}^s/\mathcal{P}^s \cong T_g\mathcal{M}^s/T_g\mathcal{P}^s &= \{[h] \mid h \in S^s_2\} \\
\text{with } [h] &= \{h + \lambda g \mid \lambda \in H^s\} \ ,
\end{aligned}$$

$D\pi_{-1}(g) : h \mapsto [h]$.

To show surjectivity means to show that given any equivalence class $[h]$, one can find a representative $h + \lambda g \in \ker DR(g)$. So we have to solve $0 = DR(g)(h + \lambda g)$. By redefining λ we may assume with no loss of generality that $tr_g h = 0$. The equation then becomes

$$-2\Delta_g\lambda + \delta_g\delta_g h + \Delta_g\lambda + \lambda = 0 \ .$$

Surjectivity of $-\Delta + id$ guarantees the existence of such a solution, injectivity guarantees its uniqueness.

So we have shown bijectivity of $D\pi_{-1}$ which ends the proof. ∎

It is easy that $\mathcal{D}^{s+1}$ acts on $\mathcal{M}^s_{-1}$, for

$$R(f^*g) = f^*R(g) = R(g) \circ f \ .$$

Thus, if $R(g) \equiv -1$, so is $R(f^*g)$.

Moreover, $\pi_{-1}(f^*g) = [f^*g] = f^*[g] = f^*\pi_{-1}(g)$, so we have

Theorem 1.6.3 $\pi_{-1} : \mathcal{M}^s_{-1} \to \mathcal{M}^s/\mathcal{P}^s$ *is $\mathcal{D}$-equivariant.*   ∎

This gives us a very nice potential model for $\mathcal{T}(M)$, namely $\mathcal{T}(M) = \mathcal{M}_{-1}/\mathcal{D}_0$.

One apparent problem is that we have worked out everything in the H^s category, not in the C^∞ category. We will deal with this by considering some $g \in C^\infty$ as $g \in H^s$ for any s and showing that certain objects constructed above as objects in H^s are indeed C^∞. This is done partially by arguing with regularity results, partially by observing that $\bigcap_{s>3} H^s = C^\infty$.

The details of this rough sketch constitute the subsequent chapter.

2 The Construction of Teichmüller Space

2.1 A Rapid Course in Geodesic Theory

The theory of geodesics began in 1697, when Johann Bernoulli challenged the mathematicians of his time with the problem of finding the shortest path between two points on a surface of revolution. The problem was solved by his brother, Jakob Bernoulli.

The first paper on geodesics appeared as early as 1732 and was written by Euler. He showed that a curve on a surface (embedded in $\mathbb{R}^3$) is a geodesic if and only if its osculating plane in any point is orthogonal to the tangent plane of the surface at the same point.

On the other hand, the fact that there always exists a shortest path between any two points on a closed and connected surface was first proved by Hilbert using calculus of variations, circa 1900.

Geodesics on a surface are locally the shortest path between two of its points, i.e. if $p = \sigma(0), q = \sigma(1)$ are close enough, $t \mapsto \sigma(t)$ is a geodesic, if and only if $\int_0^1 ||\sigma'(t)|| dt$ is minimal among all $\tilde{\sigma}$ satisfying $\tilde{\sigma}(0) = p, \tilde{\sigma}(1) = q$. This integral is invariant under reparametrization, which complicates the theory considerably. To simplify the theory, one chooses only parametrizations proportional to arc length, i.e. $||\sigma'(t)||$ constant. Then geodesics are exactly the curves satisfying the differential equation $\frac{D}{\partial t}\sigma'(t) = 0$ (where $\frac{D}{\partial t}$ denotes the covariant derivative). In local coordinates, this equation reads (for any Riemannian manifold):

$$\frac{d^2\sigma^k}{dt^2} + \Gamma_{ij}^k\left(\sigma(t)\right)\frac{d\sigma^i}{dt}\frac{d\sigma^i}{dt} = 0 \ .$$

For a surface embedded in $\mathbb{R}^3$, the covariant derivative has the following intuitive meaning: we have

$$]a, b[\overset{\sigma}{\to} S \hookrightarrow \mathbb{R}^3 \ .$$

Thus $\sigma'(t) \in T_{\sigma(t)}S \subset T_{\sigma(t)}\mathbb{R}^3 \cong \mathbb{R}^3$.

$\sigma''(t) \notin T_{\sigma(t)}S$ in general, but it is only in $\mathbb{R}^3$; to get an element of $T_{\sigma(t)}S$ "as close as possible" to $\sigma''(t)$, take the orthogonal projection.

If $\Pi(p) : \mathbb{R}^3 \to T_pS$ is the orthogonal projection, then

$$\frac{D}{\partial t}\sigma'(t) = \Pi\left(\sigma(t)\right)\sigma''(t) \ .$$

The existence and uniqueness theorem for ordinary differential equations immediately gives the following

Theorem 2.1.1 *Given any $p_0 \in M$, then there exists a neighborhood U of p_0 and $\varepsilon > 0$ such that $\forall p \in U$, $\forall v \in T_pM, ||v|| < \varepsilon$ there exists a unique geodesic $\sigma :]-2, 2[\to M$ with $\sigma(0) = p, \sigma'(0) = v$.*

Remarks 2.1.1 *If $s \mapsto \sigma(s)$ is a geodesic, so is $s \mapsto \sigma(st)$.*

If M is compact and $\partial M = \emptyset$, then geodesics can be defined for all t and all initial conditions.

Geodesics are interesting for the purposes of this lecture because the exponential map is constructed with their help:

Let γ be the geodesic of theorem 2.1.1, defined by $\gamma(0) = p, \gamma'(0) = v$. Then $\exp_p v := \gamma(1)$. By considering $\sigma : s \mapsto \gamma(st)$, which is a geodesic as well, one sees that

$$\gamma(t) = \sigma(1) = \exp_p tv \ .$$

Theorem 2.1.2 *Given p, for a sufficiently small neighborhood W of 0 in T_pM, $\exp_p : W \to M$ is a diffeomorphism onto a neighbourhood of p in M.*

PROOF: $D \exp_p(0)w = \frac{d}{dt}\exp_p(tw)\big|_{t=0} = w$, thus $D \exp_p(0) = id$. The conclusion follows from the inverse function theorem. ∎

By this theorem, the exponential map gives rise to special coordinate systems, called the *Riemann normal coordinates*. They were invented by Riemann on the occasion of his inaugural lecture and are defined as follows:

Let $X_1, \ldots, X_n$ be an orthonormal basis for T_pM. Consider the map

$$\mathbb{R}^n \supset U \to M, (a^1, \ldots, a^n) \mapsto \exp_p(a^i X_i) \ .$$

Then $(\alpha^1, \ldots, \alpha^n)$ are called the Riemann normal coordinates. The Riemann metric g on M induces a metric distance function ρ as follows: for $p, q \in M$,

$$\rho(p, q) := \inf \left\{ \int_0^1 ||\sigma'(t)|| dt \ \Big| \ \sigma : [0, 1] \to M, \sigma(0) = p, \sigma(1) = q \right\} \ .$$

Of course, $||\sigma'(t)||$ is meant to be $\sqrt{g(\sigma(t))(\sigma'(t), \sigma'(t))}$. The infimum is always achieved locally and generally, if M is a complete Riemannian manifold, i.e. if ρ is a complete metric; if σ is a geodesic and $\sigma'(0) = v$ is small enough, then for, say, $t_0 < 2 \rho(\sigma(0), \sigma(t_0)) = \int_0^{t_0} ||\sigma'(t)|| dt$.

Remark 2.1.2 *Let* $f : (M, G) \to (M, g)$ *be an isometry (i.e.* $f^*g = G$*). Let Exp, exp denote the exponential maps of* G, g *respectively. Then* $f(\mathrm{Exp}_p v) = \exp_{f(p)}(Df(p)v)$. *This is true, because isometries map geodesics to geodesics, and if* σ *is a geodesic with* $\sigma(0) = p, \sigma'(0) = v$, *then* $\gamma := f \circ \sigma$ *is a geodesic with* $\gamma(0) = f(p), \gamma'(0) = Df(p)v$.

The fact just established makes clear that an isometry is determined by its value and its derivative at some point ("by a point and a frame"). The reader will find a fine chapter on geodesics in Milnor's book on Morse theory [76].

2.2   The Free Action of $\mathcal{D}_0$ on $\mathcal{M}_{-1}$

We prove here the following result:

Theorem 2.2.1 $\mathcal{D}_0^{s+1}$ *acts freely on* $\mathcal{M}_{-1}^s$, $\mathcal{D}_0$ *acts freely on* $\mathcal{M}_{-1}$.

Remark 2.2.1 *Such a theorem is called of Bochner-Fraenkel type and such theorems hold in very general settings. A similar theorem does not hold for* $genus(M) \leq 1$.

PROOF: Suppose $f^*g = g$, $f \in \mathcal{D}_0^{s+1}$, $f \neq id$ to get a contradiction. There is a bijective, $\mathcal{D}$-equivariant correspondence between the set $\mathcal{C}$ of complex structures and $\mathcal{M}_{-1}$:

$$\mathcal{C} \xrightarrow{\cong} \mathcal{A} \xrightarrow{\cong} \mathcal{M}/\mathcal{P} \xrightarrow{\cong} \mathcal{M}_{-1} \ .$$

See theorems 1.1.1, 1.1.2, 1.3.4, 1.3.7, 1.6.2, 1.6.3.

Let $c(g)$ be the complex structure associated with g. By $\mathcal{D}$-equivariance, $f^*c(g) = c(g)$, i.e. f is holomorphic as a map from $(M, c(g))$ to itself.

We need some algebraic topology, which will be explained in detail immediately after the proof.

Lefshetz' fixed point theorem guarantees that f has a fixed point. Since f is a holomorphic map, its fixed points are isolated and therefore finitely many (the possibility $f = id$ is excluded by assumption). A fixed point z_0 of f is defined to be non-degenerate, iff $id - Df(z_0) : T_{z_0}M \to T_{z_0}M$ is an isomorphism.

Claim: All the fixed points of f are non-degenerate.

The claim follows from the fact that in local complex coordinates,

$$Df(z_0) = \begin{pmatrix} A & B \\ -B & A \end{pmatrix}$$

since f is holomorphic. Therefore, $\det(id - Df(z_0)) = (1 - A)^2 + B^2$ which is $\neq 0$ unless $A = 1$, $B = 0$. But since as an isometry f is determined completely by $f(z_0) = z_0$ and $Df(z_0) = id$, f must be the identity contrary to the assumption. This proves the claim, and moreover that for all fixed points z_0 of f, $\det(id - Df(z_0)) > 0$. As we shall soon see this implies that the Lefshetz number $\Lambda(f)$ of f is positive. On the other hand, $\Lambda(f) = \chi(M) < 0$ because $f \sim id$. This contradiction concludes the proof that $\mathcal{D}_0^{s+1}$ acts freely on $\mathcal{M}_{-1}^s$. That $\mathcal{D}_0$ acts freely on $\mathcal{M}_{-1}$ is an immediate consequence.

Note that there would be no contradiction in case of the sphere ($\chi(\text{sphere}) = 2$), and there are indeed isometries of the sphere homotopic to the identity: the whole group SO_3 of rotations of $\mathbb{R}^3$. They have indeed 2 fixed points. ∎

Let us review

The Lefshetz Fixed Point Theorem: *Given a manifold M, to any continuous $f : M \to M$, there is associated its Lefshetz number $\Lambda(f)$, also called Lefshetz fixed point index. The Lefshetz number $\Lambda(f)$ can be calculated in terms of the homology as follows:*

$f : M \to M$ induces linear maps

$$f_*^i : H_i(M, \mathbb{R}) \to H_i(M, \mathbb{R}) .$$

Then $\Lambda(f) = \sum_i (-1)^i tr \, f_*^i$.

From this, it is obvious that $\Lambda(f)$ depends only on the homotopy class of f; especially if f is homotopic to the identity, then

$$\Lambda(f) = \Lambda(id) = \sum_i (-1)^i \dim H_i(M, \mathbb{R}) = \chi(M) .$$

(The reader who is not familiar with homology can be content with the equation $\Lambda(f) = \chi(M)$ for $f \sim id$.)

In our case where M is a Riemann surface,

$$H_i(M, \mathbb{R}) = \begin{cases} \mathbb{R} & \text{for } i = 0 \text{ or } i = 2 \\ \mathbb{R}^{2 \text{ genus}(M)} & \text{for } i = 1 \\ 0 & \text{otherwise} . \end{cases}$$

Except for the torus ($\chi(torus) = 0$), the existence of a fixed point is guaranteed. (Indeed, there can be isometrics of a torus homotopic to the identity, but different from the identity, namely rotations.)

If f is differentiable, as it is in our situation, the concept of non-degeneracy is defined. If all fixed points of f are non-degenerate, there is an equivalent formula for the Lefshetz number:

$$\Lambda(f) = \sum_{z_0 \in \text{Fix}(f)} \text{sign det}(id - Df(z_0)) .$$

Once this formula is established, Lefshetz fixed point theorem is of course a trivial corollary. A reference for all of the above is [45] for a differential topology access.

The following theorem proves remark 1.4.1 after theorem 1.4.2. It is a consequence of Theorem 2.4.3.

Theorem 2.2.2 On a Riemann surface of genus$(M) > 1$, $L_{X_1}g = L_{X_2}g$ implies that $X_1 = X_2$.

PROOF: By linearity it is enough to show that $L_X g = 0$ implies $X = 0$. Vector fields X such that $L_X g = 0$ are generally called Killing fields. Let f_t be the flow generated by the vector field X, i.e. the solution of the ODE

$$\frac{d}{dt}f_t(x) = X(f_t(x)) \quad , \quad f_0(x) = x \ .$$

Differentiating the group property $f_{t+s} = f_t \circ f_s$ yields

$$\frac{d}{dt}f_t^*g = \frac{d}{ds}f_{t+s}^*g\Big|_{s=0} = f_t^*\frac{d}{ds}f_s^*g\Big|_{s=0} = f_t^*L_Xg \ .$$

Therefore, if $L_Xg = 0$, then f_t^*g is constant and equals $f_0^*g = g$.

This means, f_t is a path of isometries which are homotopic to the identity. The previous theorem implies $f_t = id$ for all t, which in turn implies

$$0 = \frac{d}{dt}f_t(p) = X\left(f_t(p)\right) = X(p) \qquad (\forall p) \ ,$$

so X vanishes everywhere. ∎

In view of theorem 1.3.2, we have proved that $\mathcal{D}_0^{s+1}$ acts freely on $\mathcal{M}^s$. It remains to show that the action is proper:

2.3 The Proper Action of $\mathcal{D}_0$ on $\mathcal{M}_{-1}$

Theorem 2.3.1 (Ebin, Palais)
$\mathcal{D}^{s+1}$ acts properly on $\mathcal{M}^s$. $\mathcal{D}$ acts properly on $\mathcal{M}$; in detail, this means: If $f_n^*g_n \xrightarrow{H^s} \hat{g}$ and $g_n \xrightarrow{H^s} g$ with $g_n \in \mathcal{M}^s, f_n \in \mathcal{D}^{s+1}$, then there exists a subsequence (f_k) of (f_n) which converges in $\mathcal{D}^{s+1}$ to some f.

PROOF: Let $f_n^*g_n =: \hat{g}_n$. Since we have assumed $s > 3$, H^s-convergence implies C^2-convergence. C^2-convergence means that the derivatives up to order 2 converge uniformly on compact sets. So, the assumptions imply $\hat{g}_n \xrightarrow{C^2} \hat{g}$, $g_n \xrightarrow{C^2} g$.

Let $\hat{e}_n, \hat{e}, e_n, e$ denote the exponential mappings at some given point (which is suppressed in the notation) with respect to $\hat{g}_n, \hat{g}, g_n, g$, respectively. Similarly denote the corresponding Christoffel symbols by $_n\hat{\Gamma}, \hat{\Gamma}, _n\Gamma, \Gamma$.

Since the latter contain derivatives of the metric, $_n\hat{\Gamma} \xrightarrow{C^1} \hat{\Gamma}$, $_n\Gamma \xrightarrow{C^1} \Gamma$. The exponential mappings are solutions of ordinary differential equations whose coefficients are the

Christoffel symbols, and as such they inherit the convergence properties of the coefficients by the general theory of ODEs. Therefore, $\hat{e}_n \xrightarrow{C^1} \hat{e}$, $e_n \xrightarrow{C^1} e$ (i.e. the convergence of the first derivatives is uniform on compact subsets of TM).

Since $\hat{e}_n$ and $\hat{e}$ are local diffeomorphisms, we also have $\hat{e}_n^{-1} \xrightarrow{C^1} \hat{e}^{-1}$ (and similarly $e_n^{-1} \xrightarrow{C^1} e^{-1}$).

A first step towards proving H^{s+1}-convergence of a subsequence (f_k) is to prove convergence of f_k and Df_k at a finite set of points on M.

Let ε ($\hat{\varepsilon}$) be small enough such that any ball of radius $< \varepsilon$ ($< \hat{\varepsilon}$) with respect to g ($\hat{g}$) is contained in a Riemann normal coordinate system. By shrinking ε and $\hat{\varepsilon}$ and choosing subsequences, one can obtain the same property with respect to all g_n ($\hat{g}_n$) since $g_n \to g$, $\hat{g}_n \to \hat{g}$.

Again using compactness, we can cover M by a finite set of neighbourhoods U_i any of which is contained in a ball of radius $< \delta$ about a point p_i. It is understood that the radius is to be $< \delta$ for any of the metrics $\hat{g}_n, g_n, \hat{g}, g$ where we suppose $\delta \leq \frac{1}{2} \min(\varepsilon, \hat{\varepsilon})$. So we can certainly use Riemann normal coordinates about p_i for U_i.

In the course of this proof, we shall use Greek letters for differential geometric indices as opposed to the indices i, n, k which denote individual members of sets and sequences.

For each of the points p_i, choose a $\hat{g}$-orthonormal basis $V_\mu^{(i)}$ ($\mu = 1, 2$) for $T_{p_i}M$:
$\hat{g}(p_i)\left(V_\lambda^{(i)}, V_\mu^{(i)}\right) = \delta_{\lambda\mu}$.

We claim that we can find a subsequence (f_k) of (f_n) such that for all i:

$$(1) \quad f_k(p_i)\to : q_i$$
$$(2) \quad Df_k(p_i)V_\mu^{(i)}\to : W_\mu^{(i)} \ .$$

The first claim is immediate from compactness. The second claim will be equally immediate once the boundedness of the sequences $Df_n(p_i)V_\mu^{(i)}$ is established.

The equation

$$g_k\left(f_k(p_i)\left(Df_k(p_i)V_\mu^{(i)}, Df_k(p_i)V_\lambda^{(i)}\right) = (f_k^* g_k)(p_i)\left(V_\mu^{(i)}, V_\lambda^{(i)}\right) \to \hat{g}(p_i)\left(V_\mu^{(i)}, V_\lambda^{(i)}\right) = \delta_{\mu\lambda}$$

shows that the left hand term is bounded by some constant C. On the other hand $g_k \to g$ uniformly on M and $f_k(p_i) \to q_i$ by the first claim. This implies that $Df_k(p_i)V_\mu^{(i)}$ are bounded. By possibly extracting further subsequences, claim 2 can be established. Doing the limit in the above formulas shows that $W_\mu^{(i)}$ is a g-orthonormal base of $T_{q_i}M$.

For any fixed U_i, write any $q \in U_i$ in Riemann normal coordinates:

$$q = \hat{e}\left(\sum_\mu a_{(i)}^\mu V_\mu^{(i)}\right) = \hat{e}_k\left(\sum_\mu a_{(i)k}^\mu V_\mu^{(i)}\right) .$$

Here, both $\hat{e}$ and $\hat{e}_k$ are to be taken at p_i. We therefore have for large k

$$\sum_\mu a_{(i)k}^\mu(q) V_\mu^{(i)} = \hat{e}_k^{-1}\hat{e}\left(\sum_\mu a_{(i)}^\mu(q) V_\mu^{(i)}\right) .$$

It immediately follows that $a_{(i)k}^\mu \xrightarrow{C^1} a_{(i)}^\mu$.

Now, for the above $q = \hat{e}\left(\sum_\mu a_{(i)}^\mu V_\mu^{(i)}\right)$, we define

$$f(q) := e\left(\sum_\mu a_{(i)}^\mu W_\mu^{(i)}\right) .$$

Then

$$f_k(q) = f_k e_k\left(\sum_\mu a_{(i)k}^\mu V_\mu^i\right) .$$

By remark 2.1.2 this is equal to

$$e_k\left(\sum_\mu a_{(i)k}^\mu Df_k(p_i) V_\mu^i\right)$$

and from what we have already said, this clearly converges to $f(q)$, and so $f_k \xrightarrow{C^1} f$ on U_i.

Since the functions f_k are functions on all of M, f is well-defined as a function on M (i.e. the definition of f on $U_i \cap U_j$ is independent of whether one uses U_i or U_j in the above formula).

We have the maps $L_i : T_{p_i}M \to T_q M$, $V_\mu^{(i)} \mapsto W_\mu^{(i)}$.

f is equal to $e \circ L_i \circ \hat{e}^{-1}$ on U_i, thus it is locally a diffeomorphism. Especially, $f(M)$ is an open subset of M. But it is also a compact subset of M, thus closed. This shows that f is surjective.

In order to show that f is also injective (and thus globally a diffeomorphism), we claim, that $f : (M, \hat{g}) \to (M, g)$ is an isometry.

This is the following calculation:

$$\hat{\rho}(p, q) = \lim \hat{\rho}_k(p, q) = \lim \rho_k\left(f_k(p), f_k(q)\right) = \rho\left(f(p), f(q)\right) .$$

The last equation holds because $\rho_k \to \rho$ uniformly on bounded sets. There is another way to prove that f is a diffeomorphism. The convergence proof given for $f_n^* g_n$, g_n works for $\hat{g}_n, (f_n^{-1})^* \hat{g}_n$ as well and shows $f_k^{-1} \xrightarrow{C^1}: f_-$. Obviously $f_- \circ f = f \circ f_- = id$, so $f_- = f^{-1}$, and f is a diffeomorphism.

The only thing which remains to be proved is that the convergence is indeed H^{s+1}-convergence. The decisive idea how to do this is due to Palais: the transformation of the Christoffel symbols under the diffeomorphisms f_k involves the second derivatives of f_k. Solving for them gives an equation that allows us to improve the convergence inductively.

Denote by $\frac{\partial x^\rho}{\partial f^\mu}$ the entries of the matrix $\left(\frac{df}{dx}\right)^{-1}$. The transformation law of the Christoffel symbols is:

$$_n\hat{\Gamma}^\lambda_{\mu\nu} = \frac{\partial x^\rho}{\partial f_k^\mu} \frac{\partial x^\sigma}{\partial f_k^\nu} \frac{\partial f_k^\lambda}{\partial x^\tau} \, _k\Gamma^\tau_{\rho\sigma} - \frac{\partial^2 f_k^\lambda}{\partial x^\sigma \partial x^\rho} \frac{\partial x^\rho}{\partial f_k^\mu} \frac{\partial x^\rho}{\partial f_k^\nu} \ . \tag{2.1}$$

Therefore,

$$\frac{\partial^2 f_k^\lambda}{\partial x^\sigma \partial x^\rho} = -\frac{\partial f_k^\mu}{\partial x^\sigma} \frac{\partial f_k^\nu}{\partial x^\rho} \, _n\hat{\Gamma}^\lambda_{\mu\nu} + \, _k\Gamma^\tau_{\rho\sigma} \frac{\partial f_k^\lambda}{\partial x^\tau} \ . \tag{2.2}$$

The Christoffel symbols converge C^1 to the corresponding symbols without the index k, the other terms on the right hand side converge C^0. Thus, the left hand side converges C^0, i.e. f_k converges C^2. Repeat the argument once to see that f_k converges C^3. Now we can switch back to H^s-convergence. Assume $f_k \xrightarrow{H^t} f$ for some $t \leq s$. This assumption is justified for $t = 3$, so start the induction here and assume $t \geq 3$. Then $_k\hat{\Gamma} \xrightarrow{H^{t-1}} \hat{\Gamma}$ and $_k\Gamma \xrightarrow{H^{t-1}} \Gamma$ (the latter convergence is even H^{s-1}, but we do not need this here). The same convergence holds for all $\frac{\partial f_k^\mu}{\partial x^\sigma}$. Since $t - 1 \geq 2$, the products on the right hand side also converge H^{t-1}, and then the equation implies $f_k \xrightarrow{H^{t+1}} f$.

We could not have started the induction at a lower level of differentiability, because the product of two H^1-convergent sequences need not converge in H^1. This is the reason why we switched to C^2-convergence in the beginning. ∎

The transformation formula 2.2 for $f, \Gamma, \hat{\Gamma}$ instead of f_k, $_k\Gamma$, $_k\hat{\Gamma}$ also permits us to improve the differentiability of f from the assumptions $f \in \mathcal{D}^{s+1}$, $g \in C^\infty$, $\hat{g} \in C^\infty$ arbitrarily often. Therefore, $f^* g \in C^\infty$ and $g \in C^\infty$ implies $f \in C^\infty$.

2.4   The Construction of Teichmüller Space

We have defined $\mathcal{T}(M) = \mathcal{M}_{-1}/\mathcal{D}_0$ as a set, and we know that $T_g \mathcal{M}^s_{-1} = \ker DR(g)$. Consider now any $g_0 \in \mathcal{M}_{-1} \subset \mathcal{M}^s_{-1}$ (for any s). Then the orbit of $\mathcal{D}_0^{s+1}$ through g_0 is a *smooth* submanifold of $\mathcal{M}^{\bullet}_{-1}$. Note that it need not be a manifold for $g \in \mathcal{M}^{\bullet}_{-1} \smallsetminus \mathcal{M}^{\bullet+1}_{-1}$, because $\mathcal{D}^{\bullet+1}_0$ acts continuously, but not smoothly on $\mathcal{M}^{\bullet}_{-1}$. But since $g_0 \in \mathcal{M}^{\bullet+1}_{-1}$ the additional derivative needed in $\frac{d}{dt}\big|_{t=0} g \circ f_t$ is available. The orbit *is* a manifold, and its tangent space at g_0 is

$$\{L_X g_0 \mid X \text{ is an } H^{\bullet} \text{ vector field}\} \ .$$

Let $h \in T_{g_0} \mathcal{M}^{\bullet}_{-1}$. Then there is the L^2-orthogonal decomposition $h = h^0 + L_X g_0$ constructed in theorem 1.4.2 for a general $h \in S^{\bullet}_2$.

Theorem 2.4.1 *If $h \in T_{g_0} \mathcal{M}^{\bullet}_{-1}$ with $g_0 \in \mathcal{M}_{-1}$, then for the decompositions $h = h^0 + L_X g_0$ according to theorem 1.4.2, not only $\delta_{g_0} h^0 = 0$, but also $tr_{g_0} h^0 = 0$.*

Remark 2.4.1 *Such tensors $h \in S^{\bullet}_2$, that satisfy $\delta_g h = 0$ and $tr_g h = 0$ are called transverse traceless (abbreviated TT). Transverse is used as a synonym to divergence free here, because the decomposition of theorem 2.4.1 shows that such tensors are indeed transversal to the orbit of $\mathcal{D}^{\bullet+1}_0$. However, the use of the word transversal is not special to this approach to Teichmüller theory, but comes from theoretical physics.*

PROOF: Since $L_X g_0 \in T_{g_0} \mathcal{M}^{\bullet}_{-1}$, we have $DR(g_0)L_X g_0 = 0$ (it is the derivative of R in a direction $L_X g_0$ along which R is constant -1).

On the other hand $DR(g_0)h = 0$, since $h \in T_{g_0} \mathcal{M}^{\bullet}_{-1}$. Therefore $DR(g_0)h^0 = 0$.

Using formula 1.5 and $\delta_{g_0} h = 0$ implies $-\Delta_{g_0}(tr_{g_0} h^0) + \frac{1}{2}(tr_{g_0} h^0) = 0$, thus $tr_{g_0} h^0 = 0$. ∎

Definition 2.4.1 $S^{TT}_2(g) := \{h \in S^{\bullet}_2 \mid \delta_g h = 0 \text{ and } tr_g h = 0\}$

Remark 2.4.2 *Consider s as suppressed in this notation for the time being. A later regularity result will show that $S^{TT}_2(g)$ does not depend on s for $g \in C^{\infty}$.*

The vectors in $S^{TT}_2(g)$ are candidates for the tangent vectors of $\mathcal{T}(M)$.

To see the (strong) implications of TT, one does the following calculations in conformal coordinates (x, y), i.e. $(g_0)_{ij} = \rho \cdot \delta_{ij}$. Let $h^{TT} = h_{11}dx^2 + 2h_{12}dx\,dy + h_{22}dy^2$. In these coordinates, traceless means $h_{11} = -h_{22} =: u$.

Thus $(v := -h_{12})$, $h^{TT} = u\, dx^2 - 2v\, dx\, dy - u\, dy^2$.

Now

$$
\begin{aligned}
0 = (\delta_{g_0} h)_i \;\; &= \;\; \frac{1}{\sqrt{g_0}} \frac{\partial}{\partial x^j} \left(h_i^j \sqrt{g_0} \right) - \frac{1}{2} h^{jk} \frac{\partial (g_0)_{jk}}{\partial x^i} \\[4pt]
&= \;\; \frac{1}{\rho} \frac{\partial}{\partial x^j} (h_{ik} g_0^{kj} \rho) - \frac{1}{2} h^{jk} \delta_{jk} \frac{\partial \rho}{\partial x^i} \\[4pt]
&= \;\; \frac{1}{\rho} \frac{\partial}{\partial x^j} (h_{ik} \delta^{kj}) - \frac{1}{2} tr_{g_0} h \frac{1}{\rho} \frac{\partial \rho}{\partial x^i} \\[4pt]
&= \;\; \frac{1}{\rho} \frac{\partial}{\partial x^k} h_{ik} \;\; .
\end{aligned}
$$

So

$$
\begin{aligned}
\frac{\partial}{\partial x} u - \frac{\partial}{\partial y} v = 0 && \text{(from } i = 1) \\[4pt]
-\frac{\partial}{\partial x} v - \frac{\partial}{\partial y} u = 0 && \text{(from } i = 2) \;\; .
\end{aligned}
$$

But these are the Cauchy-Riemann differential equations. They guarantee that $\xi(z) = u(x,y) + iv(x,y)$ is locally a holomorphic function ($z = x + iy$).

This implies that $h^{TT} = \mathrm{Re}\left((u + iv) \cdot (dx + idy)^2 \right) = \mathrm{Re}\,\xi(z) dz^2$ is a holomorphic quadratic differential.

We therefore clearly have a bijective correspondence $h^{TT} \leftrightarrow \xi(z) dz^2$ between $S_2^{TT}(g_0)$ and the space of holomorphic quadratic differentials on $(M, c(g_0))$. Moreover, since holomorphic implies C^∞ it follows that $S_2^{TT}(g_0)$ consists of only C^∞ tensors (cf. remark 2.4.2 above).

By the Riemann-Roch theorem, the space of holomorphic quadratic differentials has dimension 6 genus$(M) - 6$. This fact will be considered in more detail in the appendix.

This is the point where the present approach comes together with Teichmüller's original approach. Consider the torus case (genus$(M) = 1$) first: it is possible to map any parallelogram onto any other parallelogram conformally. This is a consequence of the well known Riemann mapping theorem. There are 3 degrees of freedom to do this according to the fact that the conformal automorphism group of the disc is of dimension 3. But it is not possible in general to map one parallelogram to another conformally in such a way that corners are mapped to corners. Unless a miracle happened, this would need 4 degrees of freedom. This is the birth of the idea of a quasiconformal mapping the definition of which we do not go into here. It relaxes the condition of conformality, at the same time trying to stay as near to conformality as possible. Let it be enough to say that two parallelograms can be mapped quasiconformally into each other preserving corners. The quasiconformal mapping "closest" to a conformal map, i.e. the unique extremal quasiconformal map is affine in this case.

The concept of quasiconformal mappings is broad enough to allow for the existence of a quasiconformal mapping between any two compact Riemann surfaces of the same genus. Again, as above there is a unique extremal quasiconformal mapping: It is not at all obvious and needed Teichmüller's brilliant insight to discover that extremal quasiconformal mappings can be identified with holomorphic quadratic differentials.

On the other hand, from the Riemannian geometer's approach, these holomorphic quadratic differentials appear quite naturally and in a straightforward way. Much of our previous work comes to fruition in the following two theorems.

Theorem 2.4.2 *Given $g_0 \in \mathcal{M}$ there exists a local C^∞ submanifold S of $\mathcal{M}^s_{-1}$ (for any s) of dimension 6 genus$(M) - 6$ passing through g_0. It contains only C^∞ metrics. Moreover, $T_{g_0} S = S^{TT}_2(g_0)$.*

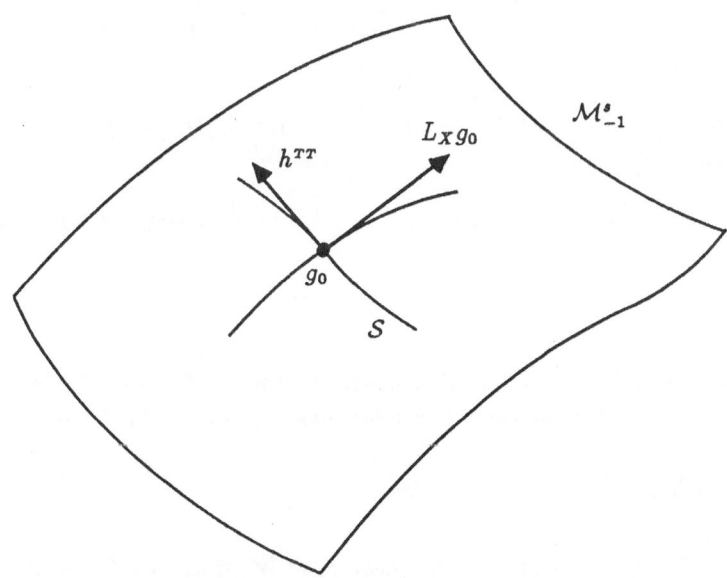

Figure 2.1: The slice

PROOF: This is essentially Poincaré's theorem:

Since g_0 is a metric, so is $g_0 + h^{TT}$ for h^{TT} small enough. By Poincaré's theorem, there exists a unique $\lambda(h^{TT})$ such that $R\left(\lambda(h^{TT}) \cdot (g_0 + h^{TT})\right) \equiv -1$.

Since $\lambda(h^{TT})$ is the unique solution of an elliptic PDE which depends smoothly on h^{TT}, $\lambda(h^{TT})$ is a smooth function of h^{TT}. λ maps some neighborhood U of 0 in S_2^{TT} into the real valued C^∞ functions on M; $\lambda(0) \equiv 1$.

Define a map

$$\Xi: \quad U \longrightarrow \mathcal{M}^s_{-1} \quad (\text{any } s)$$
$$h^{TT} \longmapsto \lambda(h^{TT}) \cdot (g_0 + h^{TT}).$$

To be specific, one can take $U = \{h^{TT} \in S_2^{TT} \mid g_0 + h^{TT} > 0\}$. We claim now that $D\Xi(0)k^{TT} = k^{TT}$. This claim implies that for U a small enough neighborhood, $\Xi(U)$ is a submanifold S (consisting only of C^∞ metrics, since g_0 and h^{TT} are C^∞ tensors and λ is a C^∞ function).

The theorem immediately follows; it only remains to prove the claim:

$$D\Xi(0)k^{TT} = \Big(\underbrace{(D\lambda(0)k^{TT})}_{=:\,\rho} \Big)g_0 + \underbrace{\lambda(0)}_{\equiv\, 1} k^{TT}.$$

It remains to show $\rho \equiv 0$.

Since $\Xi(0) = g_0$ and $R(\Xi(h^{TT})) \equiv -1$, we have $DR(g_0)D\Xi(0)k^{TT} = 0$, thus $DR(g_0)(\rho g_0 + k^{TT}) = 0$ and

$$0 = -\Delta_{g_0} tr_{g_0}/\rho g_0 + k^{TT}) + \delta_{g_0}\delta_{g_0}(\rho g_0 + k^{TT}) + \frac{1}{2}tr_{g_0}(\rho g_0 + k^{TT}) = -\Delta_{g_0}\rho + \rho \ .$$

Therefore, $\rho = 0$.   ∎

Theorem 2.4.3 *There exist neighborhoods $\tilde{W}$ of g_0 in $\mathcal{M}^s_{-1}$ and W of g_0 in S and V of id in $\mathcal{D}_0^{s+1}$ which are diffeomorphic under the mapping $\Theta: S \times \mathcal{D}_0^{s+1} \to \mathcal{M}^s_{-1}$, $\Theta(g, f) := f^* g$.*

As an immediate consequence, we have

Corollary 2.4.4 *There exists a neighbourhood $\tilde{W}$ of g_0 in $\mathcal{M}^s_{-1}$ such that every orbit $f^* g$ $(g \in \tilde{W})$ for f sufficiently close to the identity intersects S only once.*

PROOF: Consider the map $\Theta: S \times \mathcal{D}_0^{s+1} \to \mathcal{M}^s_{-1}$, $(g, f) \mapsto f^* g$. Θ is a smooth map. We want to show that

$$D\Theta(g_0, id)(h^{TT}, L_X g_0) = h^{TT} + L_X g_0 \ .$$

Since by theorem 2.4.1 every $h \in T_{g_0}\mathcal{M}_{-1}^s$ can be written as $h = h^{TT} + L_X g_0$, we see that

$D\Theta(g_0, id) : T_{g_0}\mathcal{S} \times T_{id}\mathcal{D}_0^{s+1} \to T_{g_0}\mathcal{M}_{-1}^s$ is an isomorphism. Therefore Θ is a local diffeomorphism as spelled out in the theorem. It remains to verify the formula for $D\Theta(g_0, id)$:

$$\frac{\partial}{\partial g}\Theta(g_0, id)h^{TT} = id^* h^{TT} = h^{TT}, \quad \text{since } g \mapsto f^*g \text{ is linear.}$$
$$\frac{\partial}{\partial f}\Theta(g_0, id)L_X g_0 = \frac{d}{dt}\Big|_{t=0} f_t^* g_0 = L_X g_0$$

where f_t is such that $f_0 = id, \frac{d}{dt}\Big|_{t=0} f_t = X$. ∎

Remark 2.4.3 If f and g are C^∞, so is obviously f^*g. On the other hand, every $G \in \mathcal{M}_{-1}^s$ near g_0 can be uniquely written as $G = f^*g$ with $g \in \mathcal{S}$, thus $g \in C^\infty$ and $f \in \mathcal{D}_0^{s+1}$. On the other hand, f is C^∞ if G is. This does not follow from the above theorem, but has been noted as a consequence of our proof that $\mathcal{D}^{s+1}$ acts properly on $\mathcal{M}_{-1}^s$ (see Theorem 2.3.1).

Let us repeat where we now stand:

$\mathcal{D}_0^{s+1}$ acts freely and properly on $\mathcal{M}_{-1}^s$, and $\mathcal{D}_0$ acts freely on $\mathcal{M}_{-1}$. We have decided not to work with the C^∞ topology and therefore need not discuss whether $\mathcal{D}_0$ acts properly on $\mathcal{M}_{-1}$. What we are doing is to equip $\mathcal{D}_0$ with the H^s topology for any s.

At any $g_0 \in \mathcal{M}_{-1}$, we have a slice $\mathcal{S}$ consisting of C^∞ metrics only and the local diffeomorphism $\Theta : (f, g) \mapsto f^*g$ according to theorem 2.4.3 for which, provided g is C^∞, f^*g is C^∞ if and only if f is C^∞. At this point, we get the following

Theorem 2.4.5 If the slice $\mathcal{S}$ is taken to be sufficiently small, each point of $\mathcal{S}$ corresponds to exactly one orbit of $\mathcal{D}_0$, i.e. $f^*g \in \mathcal{S} \Rightarrow f = id$.

Corollary 2.4.6 $\mathcal{T}(M) = \mathcal{M}_{-1}/\mathcal{D}_0$ is a smooth C^∞ manifold.

PROOF OF THE THEOREM: By corollary 2.4.4 the only possibility is that f^*g might be in $\mathcal{S}$ for some f far from the identity. This is excluded by properness and freeness.

Suppose that, contrary to the theorem, there are sequences f_n in $\mathcal{D}_0$ and g_n in $\mathcal{S}$ such that $g_n \xrightarrow{H^s} g_0, f_n^* g_n \xrightarrow{H^s} g_0$, all f_n outside some fixed H^{s+1} neighbourhood of id in $\mathcal{D}_0$.

Properness gives an H^{s+1} convergent subsequence (f_k) of (f_n): $f_k \to f$, and then $f^* g_0 = g_0$. Freeness implies $f = id$ contrary to the assumption that all f_n keep outside some neighbourhood of id.   ■

Remark 2.4.4 *Theorem 2.4.5 would not be true for orbits of $\mathcal{D}$ instead of orbit of $\mathcal{D}_0$.*

PROOF OF COROLLARY 2.4.6: As we have seen, $\mathcal{T}(M)$ can be locally identified with a slice S. We have to prove that these local identifications can be patched together in a compatible way. The slices S are C^∞ diffeomorphic to neighbourhoods of 0 in S_2^{TT} according to theorem 2.4.3. This means that instead of such neighbourhoods of 0 in S_2^{TT}, we can take the slices as coordinate charts.

A local coordinate chart on $\mathcal{M}_{-1}/\mathcal{D}_0$ is therefore a map θ assigning to $g \in S$ the orbit $\{f^* g \mid f \in \mathcal{D}_0\}$. Similarly let $\tilde\theta$ be another local coordinate chart. We have to show that $\tilde\theta \circ \theta^{-1}$ is C^∞ where defined. Consider the C^∞ diffeomorphism $\tilde\Theta$. Then the equation

$$\tilde\Theta(f, \tilde g) = \tilde G, \quad (f, \tilde g) \in \mathcal{D}_0 \times \tilde S \tag{2.3}$$

has unique coordinate solutions $f(\tilde G), \tilde g(\tilde G)$ which depend smoothly on $\tilde G$. Therefore

$$(\tilde\theta \circ \tilde\theta^{-1})(g) = \tilde\theta(g) = \tilde g$$

where $\tilde g$ is defined by $\tilde\Theta(f, \tilde g) = g$.

By (2.3), $\tilde\theta(g)$ depends smoothly on g and hence $\tilde\theta \circ \theta^{-1}$ is C^∞ smooth, i.e. $\mathcal{T}(M)$ is a C^∞ smooth manifold.   ■

Since we can take slices as coordinate charts, we may assume that $\theta(g) = g$ for $g \in S$.

2.5   The Principal Bundles of Teichmüller Theory

Now we have our manifolds $\mathcal{M}_{-1}$ and $\mathcal{T}(M) \cong \mathcal{M}_{-1}/\mathcal{D}_0$, and there is the obvious projection $\pi : \mathcal{M}_{-1} \to \mathcal{M}_{-1}/\mathcal{D}_0$. We should like this to be the data of a smooth $\mathcal{D}_0$ principal bundle. We could do this by working in some nice category in which the implicit function theorem holds, say the tame Fréchet category. But this is not the route we are going to take, because this category has nice properties but would require much more additional analysis and hence would delay the presentation of our main results.

Instead, we continue to equip spaces of C^∞ objects with H^s topologies and refer to explicit calculations instead of general theorems.

We have constructed $\Theta : S \times \mathcal{D}_0^{s+1} \to \mathcal{M}_{-1}^s$, $(g, f) \mapsto f^* g$ in theorem 2.4.3. For any s, $\Theta(g, \cdot) : f \mapsto f^* g$ is C^∞ because we assume g to be C^∞.

We fix s. Then Θ maps $S \times \mathcal{D}_0^{s+1}$ bijectively onto an H^s-open neighbourhood of g_0 in $\mathcal{M}_{-1}^s$. Θ has the property that it maps $U \times \mathcal{D}_0$ into $\mathcal{M}_{-1}$. The image of Θ is therefore an H^s-open neighbourhood of g_0 in $\mathcal{M}_{-1}$. We have seen that S is diffeomorphic to an open set U in $\mathbb{R}^{6\, \text{genus}(M)-6}$, and we will immediately see that $\mathcal{D}_0$ can be identified H^{s+1}-locally with the C^∞ elements of $\mathcal{X}^{s+1}(M)$. We use local coordinate charts for $\mathcal{D}_0^{s+1}$ on $\mathcal{D}_0$. They are constructed as follows: Let G be any C^∞ metric on M and Exp its exponential map. Let $f \in \mathcal{D}_0^{s+1}$.

Then $E_f : X \mapsto \mathrm{Exp}_{f(\cdot)} X (f(\cdot))$ maps a neighbourhood of 0 in the space $\mathcal{X}^{s+1}(M)$ bijectively to a neighbourhood of f in $\mathcal{D}^{s+1}$.

This is a consequence of the implicit function theorem and the fact that $DE_f(0) : X \mapsto X \circ f$ is an isomorphism.

Therefore, denoting $\mathcal{X}_f^{s+1}(M) := \{ X \circ f \mid X \in \mathcal{X}^{s+1}(M) \}$ the space of "vector fields over f", we have the isomorphisms

$$T_f \mathcal{D}_0^{s+1} \cong \mathcal{X}_f^{s+1}(M) \cong \mathcal{X}^{s+1}(M) \ .$$

Such coordinate charts preserve C^∞ if $f \in \mathcal{D}_0$: Obviously, if X is C^∞, so is $E_f(X)$. And if $E_f(X) = \mathrm{Exp}_{f(\cdot)} X (f(\cdot))$ is C^∞, then, using the fact that

$$\mathrm{Exp}_{f(p)} X (f(p)) = \widehat{\mathrm{Exp}}_p \left(Df(p)^{-1} X \right) (f(p)) = \widehat{\mathrm{Exp}}_p (f^* X)(p) \ ,$$

where $\widehat{\mathrm{Exp}}$ is the exponential map for the pulled back metric, it follows that $f^* X$ is C^∞ if $\mathrm{Exp}_{f(p)} X (f(p))$ is and thus so is X.

Thus we have a "manifold" structure on $\mathcal{D}_0$ induced by the $\mathcal{H}^{s+1}$ coordinate charts for $\mathcal{D}_0^{s+1}$. If we are willing to call $\mathcal{D}_0$ a "generalized manifold" because of this property then $\mathcal{M}_{-1}$ is also a generalized manifold with coordinate charts given by the maps Θ above, and the open sets of $\mathcal{M}_{-1}$ are generated by the image of $S \times \mathcal{D}_0$ under the map

$$\Theta : S \times \mathcal{D}_0^{s+1} \to \mathcal{M}_{-1}^s \ .$$

We shall see shortly that these are H^s open sets and that Θ is injective. Thus we have H^{s+1} topologies and H^s topologies on the spaces $\mathcal{D}_0$ and $\mathcal{M}_{-1}$ respectively.

Having constructed our coordinate charts for $\mathcal{M}_{-1}, \mathcal{D}_0$ and $\mathcal{M}_{-1}/\mathcal{D}_0$, the concept of differentiability is defined with respect to them and we can formulate

Theorem 2.5.1 $\pi : \mathcal{M}_{-1} \to \mathcal{M}_{-1}/\mathcal{D}_0$ *is* C^∞.

PROOF: $\pi\Theta(g, f) = \pi(f^*g) = g$.

By definition, π is called differentiable, iff $\pi \circ \Theta$ is. The differentiability of $(g, f) \mapsto g$ is trivial. ∎

Since we consider $\mathcal{M}_{-1}$ as the image under Θ of coordinate charts $\mathcal{D}_0 \times \mathcal{S}$, consequently

$$T_{\tilde{g}}\mathcal{M}_{-1} = D\Theta(g_0, f)(T_{g_0}\mathcal{S} \times T_f\mathcal{D}_0), \qquad \text{where}$$

$$\tilde{g} = \Theta(g_0, f) = f^*g_0, \qquad T_{g_0}\mathcal{S} \times T_f\mathcal{D}_0 \cong S_2^{TT}(g_0) \times \mathcal{X}(M)$$

$$D\Theta(g_0, f)(h^{TT}, X) = \tfrac{\partial\Theta}{\partial f}(g_0, f)X + \tfrac{\partial\Theta}{\partial g}(g_0, f)h^{TT} = L_X f^*g_0 + f^*h^{TT} = L_X\tilde{g} + \tilde{h}^{TT},$$

$$X \in \mathcal{X}(M).$$

We have used the fact that $\Theta(g_0, f) = \Theta(f^*g_0, id)$ together with the definition of the Lie derivative and the fact that $g \mapsto f^*g$ is a linear function. $\tilde{h}^{TT}$ is f^*h^{TT}. This computation essentially shows that for $\mathcal{S}$ small $D\Theta$ is always an isomorphism implying that the image of Θ is open in any H^s topology. Moreover Θ is an injective map onto its image since if $\Theta(g, f) = \Theta(\tilde{g}, \tilde{f})$ then $f^*g = \tilde{f}^*\tilde{g}$ which implies that $(\tilde{f} \circ f^{-1})^*\tilde{g} = g$ contradicting the property of the slice $\mathcal{S}$ (cf. theorem 2.4.5).

Now, since h^{TT} is trace free and divergence free with respect to g_0, $f^*h^{TT} = \tilde{h}^{TT}$ is trace free and divergence free with respect to $f^*g_0 = \tilde{g}$. Explicitly, $f^*tr_g h = tr_{f^*g} f^*h$ and $f^*\delta_g h = \delta_{f^*g} f^*h$. From this follows that f^*h^{TT} is trace free and divergence free with respect to f^*g_0 if h^{TT} is trace free and divergence free with respect to g_0.

Therefore

$$T_{\tilde{g}}\mathcal{M}_{-1} = \left\{ L_X\tilde{g} + \tilde{h}^{TT} \mid X \text{ is a } C^\infty \text{ vector field and } \tilde{h}^{TT} \text{ is } TT \text{ with respect to } \tilde{g} \right\}.$$

Formally, this is just the result we had in theorem 2.4.1 for the H^s category. But now, we have established it rigorously for the C^∞ category. Thus we have

Theorem 2.5.2 $T_g\mathcal{M}_{-1} = \{h \in C^\infty \mid DR(g)h = 0\}$.

Remark 2.5.1 *To prove that $\mathcal{M}_{-1}$ is a C^∞ manifold in the tame Fréchet category with tangent space $T_g\mathcal{M}_{-1} = \ker DR(g)$ we would have had to use a Nash-Moser type of implicit function theorem.*

Let us collect the formulas for better oversight:

Table 2.5.1

$$\pi : \mathcal{M}_{-1} \;\to\; \mathcal{M}_{-1}/\mathcal{D}_0, \qquad \cdot$$
$$\tilde{g} := \Theta(g,f) = f^*g \;\mapsto\; [g] \quad ; \; \pi \text{ is } C^\infty$$
$$D\pi(g) : T_g\mathcal{M}_{-1} \;\to\; T_{[g]}\mathcal{M}_{-1}/\mathcal{D}_0 \quad \text{is defined by} \quad D\pi \circ D\Theta = D(\pi \circ \Theta) \quad \text{or}$$

$$(D\pi \circ D\Theta)(g,f)(h^{TT}, X) = D\pi(f^*g)(L_X(f^*g) + f^*h^{TT}) = h^{TT}$$
$$\ker D\pi(\tilde{g}) = \{ L_X\tilde{g} \mid X \in \mathcal{X}(M) \} \;.$$

To get a better insight into the classical objects we are going to imitate, consider the following definitions:

Definition 2.5.1 A Banach Lie group $\mathcal{G}$ is a Banach manifold with a group structure such that the group operations are C^∞ smooth, i.e. the maps

$$\cdot : \mathcal{G} \times \mathcal{G} \to \mathcal{G}, (g_1, g_2) \mapsto g_1 \cdot g_2 \qquad \text{and} \qquad {}^{-1} : \mathcal{G} \to \mathcal{G}, g \mapsto g^{-1}$$

are C^∞ smooth.

Remark 2.5.2 *If we choose $\mathcal{G} = \mathcal{D}_0$ with its H^{s+1} topology, then $f_1 \mapsto f_1 \circ f_2$ is C^∞ smooth and so is $f_2 \mapsto f_1 \circ f_2$, but $(f_1, f_2) \mapsto f_1 \circ f_2$ is once differentiable but not C^1. Moreover $f \mapsto f^{-1}$ is not even once differentiable. The readers may wish to check these facts for themselves as an exercise. If we take $\mathcal{G} = \mathcal{D}_0^{s+1}$, the same map is only continuous. Even for $\mathcal{D}_0$, we suspect that $f \mapsto f^{-1}$ is not C^∞ smooth though we have not worked out a proof for this.*

Despite its topological drawback of not being a Banach manifold and its algebraical drawback of not being (proved to be) a smooth group, $\mathcal{D}_0$ will be able to play the same role as if it were a Banach Lie group in the following.

Definition 2.5.2 The quadruple $(\pi, \mathcal{F}, \mathcal{N}, \mathcal{B})$ is a Banach fibre bundle, iff $\pi : \mathcal{N} \to \mathcal{B}$ is a smooth map, $\mathcal{N}$ and $\mathcal{B}$ are Banach manifolds and

(i) $\pi^{-1}(x) := \mathcal{F}_x$ is a Banach manifold diffeomorphic to $\mathcal{F}$.

(ii) for any $x \in \mathcal{B}$, there exists a neighbourhood U and a diffeomorphism
$\psi : \pi^{-1}(U) \rightarrow U \times \mathcal{F}$ such that $\psi_x := \psi \mid \mathcal{F}_x$ is a diffeomorphism from $\mathcal{F}_x$ to $\mathcal{F}$.

In such a situation, ψ is called a local trivializing map.

Definition 2.5.3 $(\pi, \mathcal{G}, \mathcal{N}, \mathcal{B})$ is called a Banach principal $\mathcal{G}$-bundle, if it is a Banach fibre bundle, with $(\mathcal{G}, i)$ a Banach Lie group such that if ψ and $\tilde{\psi}$ are two local trivializing maps, then $(\psi_x \circ \tilde{\psi}_x^{-1})g = \gamma(x) \cdot g$, with some C^∞ map $\mathcal{B} \supset U \xrightarrow{\gamma} \mathcal{G}$.

Remark 2.5.3 *Analogous definitions hold, if one consistently replaces the word Banach by one of the following in definitions 2.5.1, 2.5.2, 2.5.3:*
Hilbert, Fréchet, ILH, tame Fréchet (or whatever other fancy category you may like).

The object $(\pi, \mathcal{M}_{-1}, \mathcal{M}_{-1}/\mathcal{D}_0)$ we are considering has formally all the data to make a principal bundle but falls short of the technical axioms in some respect. We will call it a weak principal bundle for this; this term will not be defined technically but is simply meant to invite the reader to consider the formal analogy with a principal bundle. What we have from the definition of a principal bundle is the following:

(1) $\pi : \mathcal{M}_{-1} \rightarrow \mathcal{M}_{-1}/\mathcal{D}_0$ is C^∞ smooth with respect to the H^s topology on $\mathcal{M}_{-1}$

(2) We can find a neighbourhood $U \subset \mathcal{T}(M)$ around any point of $\mathcal{T}(M)$ such that $\pi^{-1}U \cong U \times \mathcal{D}_0$. Namely, the diffeomorphism is $\pi^{-1}U \xleftarrow{\Theta} U \times \mathcal{D}_0$. We let $\Psi := \Theta^{-1}$.

(3) Ψ maps $\pi^{-1}[g]$ onto $[g] \times \mathcal{D}_0$

(4) For $\Psi, \tilde{\Psi}$ belonging to $U, \tilde{U}$ (and $[g] \in U \cap \tilde{U}$), we have $\Psi_{[g]} \tilde{\Psi}_{[g]}^{-1} f = \varphi([g]) \circ f$ where $\varphi : [g] \mapsto \varphi([g]), \mathcal{T}(M) \rightarrow \mathcal{D}_0$ is a smooth map.

What we lose from the definition of a principal bundle is:

(1) $\mathcal{M}_{-1}$ is not a manifold in the traditional sense

(2) $\mathcal{D}_0$ is not viewed as a Lie group in the traditional sense.

Concerning the last point, let us repeat that for the composition of diffeomorphisms

$f \mapsto f \circ g$ is C^∞ in any topology,

$g \mapsto f \circ g$ is C^∞ in the H^s topology, provided f is C^∞,

$(f, g) \mapsto f \circ g$ is not even C^1 in the H^s topology although it is once differentiable. Moreover $f \mapsto f^{-1}$ is not even once differentiable.

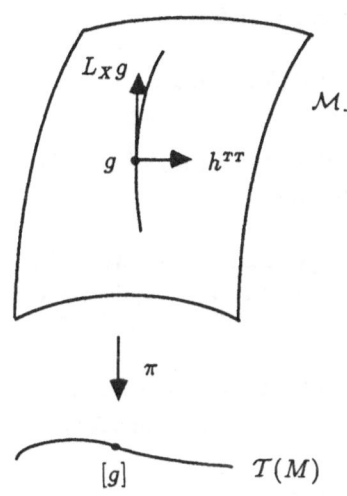

Figure 2.2

In $T_g\mathcal{M}_{-1} = \{h \mid h = L_X g + h^{TT}\}$ we can distinguish vertical vectors (i.e. those tangent to the fibre), namely those of the form $L_X g$. In general principal bundles, vertical vectors are those in $\ker D\pi$ where π is the projection from the total space to the base space. But in our case, we can also distinguish horizontal vectors in a natural way, namely those of the form h^{TT}. A natural choice of horizontal vectors is more structure than one has in a general principal bundle.
We have an isomorphism

$$D\pi(g) : S_2^{TT}(g) \xrightarrow{\cong} T_{[g]}\mathcal{T}(M) .$$

Let us introduce yet another model for $\mathcal{T}(M)$. Since $\mathcal{T}(M) \cong \mathcal{M}_{-1}/\mathcal{D}_0$ and $\mathcal{M}_{-1} \cong \mathcal{A}$, we hope that $\mathcal{T}(M) \cong \mathcal{A}/\mathcal{D}_0$ and that we get a weak principal bundle $(\pi, \mathcal{D}_0, \mathcal{A}, \mathcal{A}/\mathcal{D}_0)$. This will be made precise now.

Remember that $\mathcal{A} = \{J \mid J^2 = -id\}$.

$\mathcal{D}_0$ operates on $\mathcal{A}$ by $(J, f) \mapsto f^*J$. Formally, the tangent space to the orbit is: $T_J(\mathcal{D}_0^*J) = \left\{ \frac{d}{dt}(f_t^*J)\big|_{t=0} \right\} = \{L_X J\}$ (where $f_0 = id$, $\frac{df_t}{dt}\big|_{t=0} = X$).

The Lie derivative of a 1-1 tensor H in coordinates is:

$$(L_X H)^i{}_j = \frac{\partial H^i{}_j}{\partial x^k} X^k + H^i{}_k \frac{\partial X^k}{\partial x^j} - H^k{}_j \frac{\partial X^i}{\partial x^k} .$$

Working rigorously, we choose some $J \in \mathcal{A} \subset \mathcal{A}^s$ and let $\mathcal{D}_0^{s+1}$ operate on $\mathcal{A}^s$. Then the tangent space at J to the orbit through J is $\{L_X J \mid X \in \mathcal{X}^{s+1}\}$.

There is a $\mathcal{D}$-invariant L^2-metric on $\mathcal{A}^\bullet$ or $\mathcal{A}$. It is defined as follows: let $H, K \in T_J\mathcal{A}^\bullet$. This means $HJ = -JH$, $KJ = -JK$ according to theorem 1.1.3. Define

$$\langle\langle H, K\rangle\rangle_J := \int_M tr(HK)d\mu_{g(J)} \ . \tag{2.4}$$

Here, $g(J)$ is the metric of constant curvature -1 which is associated to J:
$g(J) = (\Phi \circ \pi_{-1})^{-1}(J)$. We will denote the map $(\Phi \circ \pi_{-1})^{-1}$ by $\tilde{\Psi}$ in the following. This definition is in complete analogy to the L^2-metric on $\mathcal{M}_{-1}$: for

$$h, k \in T_g\mathcal{M}_{-1}, \quad \langle\langle h, k\rangle\rangle_g := \int_M tr(h^\sharp, k^\sharp)d\mu_g \ .$$

Here and in the following, we denote by the superscript $\sharp$ the raising of an index; similarly, we will denote the lowering of an index by the subscript $\flat$.

Contrary to possible expectations, the natural $\mathcal{D}$-equivariant map between $\mathcal{A}$ and $\mathcal{M}_{-1}$ is *not* an isometry, if $\mathcal{A}$ and $\mathcal{M}_{-1}$ are equipped with their natural $\mathcal{D}$-equivariant L^2-metrics given above. We will see this soon.

Given $J \in \mathcal{A}$, we can define a map $\alpha_J : \mathcal{X}(M) \to C^\infty(T_1^1(M))$ by $\alpha_J(X) := L_X J$. It has an L^2-adjoint α_J^*, and $\alpha_J^*\alpha_J$ is a second order elliptic operator defined on $\mathcal{X}(M) \cong T_J\mathcal{A}$. We can decompose

$$T_J\mathcal{A} = \text{range } \alpha_J \oplus \ker \alpha_J^* \ .$$

If A is a g-symmetric trace free 1-1 tensor, then a straightforward computation shows that in conformal coordinates $g_{ij} = p \cdot \delta_{ij}$, we have the following formula for $\alpha_J^*(A)$:

$$\begin{array}{rcl} \alpha_J^*(A)^1 & = & -\frac{2}{p}(\delta_{g(J)}A)_2 \\ \alpha_J^*(A)^2 & = & +\frac{2}{p}(\delta_{g(J)}A)_1 \end{array} \tag{2.5}$$

The following theorem is in complete analogy to the decomposition

$$T_g\mathcal{M}_{-1}^\bullet = \{h \mid h = L_X g + h^{TT}\} \ .$$

Theorem 2.5.3 *Consider the manifold $\mathcal{A}^\bullet$, and let $J \in \mathcal{A}$. Then*

$$T_J\mathcal{A}^\bullet = \{H \mid H = L_X J + H^{TT}\} \ .$$

H^{TT} *is divergence free with respect to* $g(J)$, *and the decomposition* $H = L_X J + H^{TT}$ *is* L^2-*orthogonal.*

Remark 2.5.4 *We do not lay any stress on the fact that $H^{\tau\tau}$ is trace free, because any $H \in T_J \mathcal{A}^s$ is trace free.*

Up to now everything we have done with $\mathcal{A}$ is in complete analogy to what we did earlier with $\mathcal{M}_{-1}$. The difference is that we have no analog for Poincaré's theorem which would select a unique almost complex structure from a whole class of more general objects as we could select a unique metric of curvature -1 form a conformal class of metrics with any curvature. Remember that we used Poincaré's theorem to construct our slice $\mathcal{S}$. This necessity of constructing a slice is the reason why we introduced $\mathcal{M}_{-1}$. But now, having constructed this slice with the help of $\mathcal{M}$ and $\mathcal{M}_{-1}$, we can map it back into $\mathcal{A}$ as follows: Recall that we have the maps $\Xi : \mathcal{M}^s \to \mathcal{M}^s_{-1}$ where $\Xi[g] := \lambda(g) \cdot g$ with λ being defined by $R(\lambda(g) \cdot g) \equiv -1$ and $\Phi : \mathcal{M}/\mathcal{P} \to \mathcal{A}$ (cf. theorem 1.3.4). Then if $\Psi := \Phi^{-1}$, the map $\tilde{\Psi} := \Xi \circ \Psi$ is a diffeomorphism from $\mathcal{A}^s$ to $\mathcal{M}^s_{-1}$ that preserves C^∞. So, instead of $\mathcal{S}$ we can now use $\tilde{\Psi}^{-1}(\mathcal{S})$. In what follows, we shall not distinguish between $\mathcal{S}$ and $\tilde{\Psi}^{-1}\mathcal{S}$ and speak of $\mathcal{S} \subset \mathcal{A}$ instead of $\tilde{\Psi}^{-1}\mathcal{S} \subset \mathcal{A}$.

Again, we have our map $\Theta : \mathcal{S} \times \mathcal{D}_0^{s+1} \to \mathcal{A}^s$, $\Theta(J, f) = f^*J$.

As before, we get a weak principal $\mathcal{D}_0$ bundle $\pi : \mathcal{A} \to \mathcal{A}/\mathcal{D}_0$ with the C^∞ manifold structures on $\mathcal{A}$ and $\mathcal{A}/\mathcal{D}_0$ defined as before with respect to the H^s topology.

Let us understand the map $\tilde{\Psi}$ more thoroughly. $\Phi[g] := -g^{-1}\mu_g$. We have seen there, that we can identify $T_{[g]}\mathcal{M}^s/\mathcal{P}^s \cong S_2^s(g)^\tau$ and that $D\Phi[g]h = -HJ$ with $J = \Phi[g], H = h^\sharp$ for all $h \in T_{[g]}\mathcal{M}^s/\mathcal{P}^s$. It follows that

$$D\Psi(J)H = (HJ)_\flat$$

as one easily checks:

$$h \stackrel{?}{=} \left(D\Psi(J) \circ D\Phi[g]\right)h = D\Psi(J)(-HJ) = (-HJ^2)_\flat = H_\flat = h \ .$$

$D\Xi[g]h = (D\lambda(g)h) \cdot g + \lambda(g) \cdot h =: \rho \cdot g + h$ where we have chosen the representative $g \in \mathcal{M}^s_{-1}$ for $[g]$. ρ can be calculated by the condition that $\rho g + h \in T_g\mathcal{M}^s_{-1}$, i.e. $DR(g)(\rho g + h) = 0$. Using formula 1.5 in chapter 2.6 this reduces to $\Delta\rho - \rho = \delta_g\delta_g h$. The chain rule gives us

$$D\tilde{\Psi}(J)H = (D\Xi \circ D\Psi)(J)H = \rho g + h$$

$$\text{where} \qquad h = (HJ)_\flat, \quad g = \tilde{\Psi}(J), \quad \rho = (\Delta - 1)^{-1}\delta_g\delta_g h \ . \tag{2.6}$$

We can now verify our claim from above:

Theorem 2.5.4 *The $\mathcal{D}$-equivariant map $\tilde{\Psi} : \mathcal{A} \to \mathcal{M}^s_{-1}$ is not an isometry if $\mathcal{A}^s$ and $\mathcal{M}^s_{-1}$ are given the natural $\mathcal{D}$-invariant L^2-metrics.*

PROOF: Let $g = \tilde{\Psi}(J)$. For $\tilde{\Psi}$ to be an isometry, it is necessary and sufficient that

$$\langle\langle D\tilde{\Psi}(J)H, D\tilde{\Psi}(J)K\rangle\rangle_g = \langle\langle H, K\rangle\rangle_J .$$

The right hand side is $\int tr\, HK\, d\mu_g$. The left hand side is

$$\begin{aligned}
\int_M tr\left[(\rho(h)g + h)^\flat (\rho(k)g + k)^\flat\right] d\mu_g &= \int_M tr\left[(\rho(h) \cdot I + HJ)(\rho(h) \cdot I + HK)\right] d\mu_g \\
&= \int_M (tr\, \rho(h)\rho(k)I + tr\, HJKJ)\, d\mu_g \\
&= 2\int_M \rho(h)\rho(k)d\mu_g + \int_M tr\, HK\, d\mu_g .
\end{aligned}$$

So $\tilde{\Psi}$ is not an isometry unless $\rho \equiv 0$ which by (2.6) only happens if $\delta_g\delta_g h = 0$. ∎

Corollary 2.5.5 *If H is divergence free, then so is $D\tilde{\Psi}(J)H$.*

PROOF: $D\tilde{\Psi}(J)H = \rho(h)g + h$ with $h = (HJ)_\flat$.

It is an exercise to calculate explicitly that HJ is divergence free, if H is. But we can argue this point in words, too. A 1-1 tensor H is divergence free if and only if $H_\flat$ is the real part of a holomorphic quadratic differential. Multiplication of a divergence free 1-1 tensor by J corresponds to multiplication by i of the associated holomorphic quadratic differential. This operation preserves holomorphy.

Therefore, $\Delta\rho(h) - \rho(h) = \delta_g\delta_g h = 0$, hence $\rho(h) = 0$, and $\rho(h)g + h = h$ is divergence free. ∎

The first part of the following theorem is immediate now:

Theorem 2.5.6 *$D\tilde{\Psi}$ takes horizontal vectors to horizontal vectors and vertical vectors to vertical vectors.*

PROOF: A horizontal vector on $T_g\mathcal{M}_{-1}$ or $T_J\mathcal{A}$ means a transverse traceless vector, hence the first part of the theorem. The second part is obvious from $\mathcal{D}$-equivariance:

$$f_t^* \left(\tilde{\Psi}(J)\right) = \tilde{\Psi}(f_t^* J) .$$

Differentiation with respect to t at $t = 0$ yields

$$L_X \tilde{\Psi}(J) = D\tilde{\Psi}(J) L_X J \quad .$$

∎

Surprisingly, we can put another Riemannian structure on $\mathcal{M}^s_{-1}$ which makes $\tilde{\Psi}$ an isometry. Unlike the L^2-metric, it is non-degenerate on $\mathcal{M}_{-1}$, but not on $\mathcal{M}^s$. It is defined by

$$\langle\!\langle\!\langle h, k \rangle\!\rangle\!\rangle_g = \int\limits_M tr(H^T K^T) d\mu_g \quad \text{for } h, k \in T_g\mathcal{M}^s_{-1}$$

where $H^T = H - \left(\frac{1}{2} tr\, H\right) \cdot I$ is the trace free part of H and $H = h^\sharp$, and similarly for K.

All properties of a Riemannian metric are obvious except for non-degeneracy. Thus let $\langle\!\langle\!\langle h, k \rangle\!\rangle\!\rangle_g = 0$. Then $H^T = 0$, hence $H = \lambda I$, $h = \lambda g$. Using $h \in T_g\mathcal{M}^s_{-1} = \ker DR(g)$, we get $-\Delta\lambda + \lambda = 0$, hence $\lambda = 0$ from formula (1.5).

Reconsidering the proof of theorem 2.5.4 and recalling that $tr\, HJ = tr\, HK = 0$, we see that taking the trace free part simply kills the terms which contain $\rho(h)$ or $\rho(k)$. Therefore we get

$$\langle\!\langle\!\langle D\tilde{\Psi}(J)H, D\tilde{\Psi}(J)K \rangle\!\rangle\!\rangle_g = \langle\!\langle H, K \rangle\!\rangle_J \quad .$$

Since the trace commutes with pullbacks, $\langle\!\langle\!\langle \cdot, \cdot \rangle\!\rangle\!\rangle$ is $\mathcal{D}$-equivariant, too.

Since $\langle\!\langle \cdot, \cdot \rangle\!\rangle$ and $\langle\!\langle\!\langle \cdot, \cdot \rangle\!\rangle\!\rangle$ are $\mathcal{D}$-equivariant, they pass to the quotient $\mathcal{T}(M) \cong \mathcal{M}_{-1}/\mathcal{D}_0 \cong \mathcal{A}/\mathcal{D}_0$. We shall discuss this metric in the next section.

We know that every $H \in T_J\mathcal{A}$ can be decomposed uniquely into a horizontal and a vertical component: $H = H^{TT} + L_X J$. Therefore, given $H \in T_{[J]}\mathcal{T}(M)$, for any $J \in \pi^{-1}[J]$, there exists a unique horizontal vector $\tilde{H} \in T_J\mathcal{A}$ such that $D\pi(J)\tilde{H} = H$. It is called the horizontal lift of H (to J). Clearly, one can take the horizontal lift of a whole vector field: given a vector field X on $\mathcal{T}(M)$, $\tilde{X}$ is the vector field on $\mathcal{A}$, determined by the condition that $\tilde{X}(J)$ is the unique horizontal vector at J such that $D\pi(J)\tilde{X}(J) = X([J])$. We get the following

Theorem 2.5.7 *The horizontal lift of a vector field X on $\mathcal{T}(M)$ is a $\mathcal{D}$-invariant vector field on $\mathcal{A}$ (or on $\mathcal{M}_{-1}$ if we choose this a model).*

PROOF: Let $X : \mathcal{T}(M) \to T\mathcal{T}(M)$ be a vector field on $\mathcal{T}(M)$ and $\tilde{X} : \mathcal{A} \to T\mathcal{A}$ its horizontal lift. Let $O_f(J) := f^*J$ for $f \in \mathcal{D}$. The map $J \mapsto O_f(J)$ is linear, hence $\mathcal{D}_J(O_f(J))H =$

$O_f(H)$. Moreover $J \mapsto O_f(J)$ is a diffeomorphism of $\mathcal{A}$. We have to show the $\mathcal{D}$-invariance of $\tilde{X}$, i.e. $(D\, O_f^{-1})\tilde{X}\,(O_f(J)) = \tilde{X}(J)$. This amounts to checking two things.

(1) $\mathcal{D}\pi\left(O_f^{-1}\left(\tilde{X}\,(O_f(J))\right)\right) = X(J)$

(2) $O_f^{-1}\left(\tilde{X}\,(O_f(J))\right)$ is horizontal

The first is easy: let $J' := O_f(J)$; then

$$\mathcal{D}\pi\left(O_f^{-1}\tilde{X}(J')\right) = \mathcal{D}(\pi \circ O_f^{-1})\tilde{X}(J') = \mathcal{D}\pi\tilde{X}(J') = X\left([J]\right) \ .$$

The second claim is equivalent to showing that

$$\left\langle\!\left\langle O_f^{-1}\tilde{X}(J'), L_\beta J\right\rangle\!\right\rangle_J = 0 \qquad \text{for all } L_\beta J \ .$$

We use the $\mathcal{D}$-invariance of $\langle\!\langle\cdot,\cdot\rangle\!\rangle$ and the following

Lemma 2.5.8 $f^* L_\beta J = L_{f^*\beta} f^* J.$

Then $\left\langle\!\left\langle O_f^{-1}\tilde{X}(J'), L_\beta J\right\rangle\!\right\rangle_J = \left\langle\!\left\langle \tilde{X}(J'), f^*(L_\beta J)\right\rangle\!\right\rangle_{f^*J} = \left\langle\!\left\langle \tilde{X}(J'), L_{f^*\beta} J'\right\rangle\!\right\rangle_{J'} = 0$ *since* $\tilde{X}(J')$ *is horizontal and* $L_{f^*\beta}J'$ *is vertical.* ∎

PROOF OF THE LEMMA   Choose h_t such that $h_0 = id$, $\frac{dh_t}{dt}\big|_{t=0} = \beta$. Then
$$f^*(L_\beta J) = f^*\left(\frac{d}{dt}h_t^* J\right)\Big|_{t=0} = \frac{d}{dt}f^* h_t^* J\Big|_{t=0} = \frac{d}{dt}(f^{-1}\circ h_t\circ f)^*(f^*J)\Big|_{t=0} = L_{f^*\beta}f^*J \ . \qquad ∎$$

2.6   The Weil-Petersson Metric on $\mathcal{T}(M)$

There have been several metrics introduced on $\mathcal{T}(M)$. Teichmüller himself introduced a metric on $\mathcal{T}(M)$ whose definition made use of quasiconformal mappings. Later, Kobayashi introduced a metric which depends heavily on the complex structure on $\mathcal{T}(M)$. Royden [94] proved, that both metrics are the same. The Teichmüller-Kobayashi metric is not a Riemannian metric but a Finsler metric. It is complete but we shall not discuss this metric here. We are going to construct the induced L^2-metric on $\mathcal{T}(M)$, a Riemannian metric on $\mathcal{T}(M)$. It turns out that it is the same (up to a factor) as the Weil-Petersson metric. This will constitute the contents of the present section.

Scott Wolpert [126] proved that the Weil-Petersson metric is not complete and because of this it has been considered by some to be of minor importance. We argue however that it *is* interesting, and that the lack of completeness does have significance: it means there exist geodesics which cannot be continued forever. As we shall later see (section 5.5), a point beyond which a geodesic cannot be continued corresponds to the collapsing of a handle of a Riemann surface, i.e. the geodesic "crashes in finite time into a Riemann surface of lower genus".

The L^2-metric on $\mathcal{A}$ induces a metric on $T(M) = \mathcal{A}/\mathcal{D}_0$ as follows: given vector fields X, Y on $T(M)$, let $\tilde{X}, \tilde{Y}$ be their horizontal lifts. For any $[J] \in T(M)$, we choose any $J \in \pi^{-1}[J] \subset \mathcal{A}$ and define $\langle X[J], Y[J] \rangle_{[J]} := \langle\langle \tilde{X}(J), \tilde{Y}(J) \rangle\rangle_J$.

By $\mathcal{D}$-invariance, this definition is independent of the choice of J. We can do the same thing with $\mathcal{M}_{-1}$ instead of $\mathcal{A}$. Since $\tilde{\Psi} : (\mathcal{A}, \langle\langle \cdot, \cdot \rangle\rangle) \to (\mathcal{M}_{-1}, \langle\langle\langle \cdot, \cdot \rangle\rangle\rangle)$ is a $\mathcal{D}$-equivariant isometry, $\langle\langle\langle \cdot, \cdot \rangle\rangle\rangle$ on $\mathcal{M}_{-1}$ induces the same metric $\langle \cdot, \cdot \rangle$ on $T(M)$. Note that the induced L^2-metric on $T(M)$ was defined via *horizontal* lifts. Therefore in the case of $\pi : \mathcal{M}_{-1} \to T(M)$ it does not make any difference whether $\mathcal{M}_{-1}$ is equipped with $\langle\langle \cdot, \cdot \rangle\rangle$ or $\langle\langle\langle \cdot, \cdot \rangle\rangle\rangle$ since these metrics agree for horizontal vector fields; the difference in choice of the metric on $\mathcal{M}_{-1}$ disappears in the quotient.

In the context of number theory, Petersson introduced an inner product on the spaces of modular forms of arbitrary weight. We shall restrict ourselves to the case of weight 2, because modular forms of weight 2 are nothing else than holomorphic quadratic differentials. That this inner product should give a metric on Teichmüller space was remarked by André Weil in a letter to Lars Ahlfors.

Let $\xi dz^2, \gamma dz^2$ be two holomorphic quadratic differentials on $(M, g) = (M, c(g))$. Introduce conformal (in other words: complex) coordinates, i.e. $g_{ij} = \lambda \delta_{ij}$, thus $g_{ij} dx^i dx^j = \lambda |dz|^2$. Then

$$\frac{\xi dz^2 \cdot \overline{\gamma dz^2}}{(\lambda |dz|^2)^2} = \frac{\xi \bar{\gamma}}{\lambda^2}$$

is a function (i.e. of weight 0) and can therefore be integrated with respect to the Riemannian volume element $\lambda dx \, dy$. We define:

$$\langle \xi, \gamma \rangle_{WP} := \text{Re} \int_M \frac{\xi \bar{\gamma}}{\lambda} dx \, dy \tag{2.7}$$

the Weil-Petersson metric on $T(M)$.

It has been shown by Ahlfors [5] that the holomorphic sectional curvature and the Ricci curvature of $T(M)$ with respect to the WP metric are negative, and as already noted that the WP metric is not complete.

Theorem 2.6.1 $\langle \cdot, \cdot \rangle_{WP} = \frac{1}{2} \langle \cdot, \cdot \rangle$.

PROOF: Let

$$\begin{aligned} h^{TT} &= u\, dx^2 - 2v\, dx\, dy - u\, dy^2 = \mathrm{Re}\,((u+iv)dz^2) \\ k^{TT} &= \mathrm{Re}\,((u'+iv')dz^2) \ . \end{aligned}$$

Let $u + iv =: \xi$, $u' + iv' =: \gamma$ and $D\pi h^{TT} =: h$, $D\pi k^{TT} = k$.

Then $\langle h, k \rangle_{[g]} = \langle\langle H^{TT}, K^{TT} \rangle\rangle_J$.

A calculation in conformal coordinates gives:

$$H^{TT} = \frac{1}{\lambda} \begin{bmatrix} u & -v \\ -v & -u \end{bmatrix}, K^{TT} = \frac{1}{\lambda} \begin{bmatrix} u' & -v' \\ -v' & -u' \end{bmatrix}, H^{TT} K^{TT} = \frac{1}{\lambda^2} \begin{bmatrix} uu'+vv' & * \\ * & uu'+vv' \end{bmatrix} ,$$

$$\int\limits_M tr(H^{TT}K^{TT})d\mu_g = 2 \int \frac{uu'+vv'}{\lambda} dx\, dy \ .$$

Since $\mathrm{Re}\,(\xi\bar{\gamma}) = \mathrm{Re}\,((u+iv)(u'-iv')) = uu' + vv'$, the right hand side is $2\,\langle \xi, \gamma \rangle_{WP}$. ∎

3 $\mathcal{T}(M)$ is a Cell

3.1 Dirichlet's Energy on Teichmüller Space

On our way towards a proof of the claim in the headline we are going to make use of Dirichlet's energy. Dirichlet asserted the following well known principle: In order to solve the "harmonic" equation

$$\Delta u = 0 \quad \text{in } \Omega, \quad u\big|_{\partial\Omega} = \varphi$$

it suffices to minimize the "energy"

$$E(u) = \frac{1}{2} \int_\Omega \nabla u \nabla u = \frac{1}{2} \int_\Omega \left[\left(\frac{\partial u}{\partial x} \right)^2 + \left(\frac{\partial u}{\partial y} \right) \right] dx \, dy$$

among all functions $u : \Omega \to \mathbb{R}$ satisfying $u\big|_{\partial\Omega} = \varphi$.

A similar principle is known in the theory of minimal surfaces: A minimal surface of the type of the disc Ω is a map $u : \Omega \to \mathbb{R}^3$ such that $u\big|_{\partial\Omega} : \partial\Omega \to \Gamma$ is a homeomorphism (where Γ is a given closed curve in $\mathbb{R}^3$). The critical points of

$$E(u) := \frac{1}{2} \sum_j \int \nabla u^j \cdot \nabla u^j \qquad (\text{where } u = (u^1, u^2, u^3))$$

are minimal surfaces spanning Γ.

We want to extend the concept of Dirichlet's energy to Teichmüller's moduli space.

Let (M, g) be a compact oriented surface, $\partial M = \emptyset$; $R(g)$ need not be $\equiv -1$. Let (N, G) be a Riemannian n-manifold, which we may assume for ease of exposition to be embedded in $\mathbb{R}^K$ for some K.

We consider maps $S : M \to N \subset \mathbb{R}^K$, and its components in $\mathbb{R}^K$. $S = (S^1, \ldots, S^K)$.

Dirichlet's energy of S is defined in its extrinsic form as follows:

$$E(g, S) := E_g(S) := \frac{1}{2} \int\limits_M g(x)(\nabla_g S^\alpha, \nabla_g S^\alpha) d\mu_g = \frac{1}{2} \int\limits_M g^{ij} \frac{\partial S^\alpha}{\partial x^i} \frac{\partial S^\alpha}{\partial x^j} d\mu_g .$$

The term under the integral sign excluding $d\mu_g$ is called the energy density. The summation convention is meant to apply to α, too, for we need not distinguish upper and lower indices in euclidean space.

Using local coordinates on N as well, one gets the intrinsic form of Dirichlet's energy:

$$E(g, S) = \frac{1}{2} \int\limits_M G_{\alpha\beta} g^{ij} \frac{\partial S^\alpha}{\partial x^i} \frac{\partial S^\beta}{\partial x^j} d\mu_g .$$

It will be clear from the context or explicitly stated whether α runs from 1 to K or from 1 to n.

Note also that Dirichlet's energy depends on the metric G as well as on g. But we suppress G in the notation considering it as fixed.

Using complex notations on M ($z = x + iy$, $\bar{z} = x - iy$, $2\partial_z := \partial_x - i\partial_y$, $2\partial_{\bar{z}} := \partial_x + i\partial_y$), we get

$$E(g, S) = 2 \int\limits_M G_{\alpha\beta} S_z^\alpha S_{\bar{z}}^\beta dx\, dy .$$

If dim $N = 2$, $G_{\alpha\beta} = \rho \delta_{\alpha\beta}$ one can use complex variables in the image, too, thus getting

$$E(g, S) = \int\limits_M \rho\left(S, \bar{S}\right) \left(|S_z|^2 + |S_{\bar{z}}|^2\right) dx\, dy ,$$

where $S = S^1 + iS^2$, $\bar{S} = S^1 - iS^2$ and $|S_z|^2 = S_z \overline{S_z} = S_z(\bar{S})_{\bar{z}}$.

Critical points of the map $S \mapsto E_g(S)$ are called *harmonic maps*. In the context of Riemannian manifolds, this concept was introduced by Salomon Bochner.

Let us consider the Euler-Lagrange equations for Dirichlet's energy. In the extrinsic form, let $p \in N$ and note that $T_p N \subset \mathbb{R}^K$. Let $\Pi(p) : \mathbb{R}^K \to T_p N$ be the orthogonal projection.

Then, S is harmonic if and only if

$$\Delta S := \Pi\left(S(x)\right)\left(\Delta_g S\right)(x) = 0 .$$

Here, $\Delta_g S = (\Delta_g S^1, \ldots, \Delta_g S^K)$, and Δ_g is the Laplace-Beltrami operator on (M, g). Using the covariant derivative, the equation can be written as

$$\frac{D}{\partial x}\frac{\partial S}{\partial x} + \frac{D}{\partial y}\frac{\partial S}{\partial y} = 0$$

or, using the symbols $\nabla_{\frac{\partial}{\partial x}}$ for $\frac{D}{\partial x}$ and $\nabla_{\frac{\partial}{\partial y}}$ for $\frac{D}{\partial y}$, we can rewrite this as

$$\nabla_{\frac{\partial}{\partial x}}\frac{\partial S}{\partial x} + \nabla_{\frac{\partial}{\partial y}}\frac{\partial S}{\partial y} = 0 \ .$$

In the intrinsic form, $S = (S^1, \ldots, S^n)$ in local coordinates on N, and S is harmonic iff

$$\Delta S = \frac{1}{\sqrt{g}}\frac{\partial}{\partial x^i}\left(g^{ij}\sqrt{g}\frac{\partial}{\partial x^j}S^\alpha\right) + \Gamma^\alpha_{\gamma\beta}\frac{\partial S^\gamma}{\partial x^i}\frac{\partial S^\beta}{\partial x^j}g^{ij} = 0 \ .$$

Again, when $\dim N = 2$, we can use complex coordinates on M and N, and we get

$$\rho_S S_z S_{\bar z} + \rho S_{z\bar z} = 0 \ . \tag{3.1}$$

If we define covariant differentiation operators $\nabla_{\frac{\partial}{\partial z}}$ and $\nabla_{\frac{\partial}{\partial \bar z}}$ in conformal coordinates on (M, g_0) by

$$\nabla_{\frac{\partial}{\partial z}} = \frac{1}{2}\left(\nabla_{\frac{\partial}{\partial x}} - i\nabla_{\frac{\partial}{\partial y}}\right) \quad \text{and} \quad \nabla_{\frac{\partial}{\partial \bar z}} = \frac{1}{2}\left(\nabla_{\frac{\partial}{\partial x}} + i\nabla_{\frac{\partial}{\partial y}}\right) \ ,$$

then the condition for harmonicity (3.1) can be rewritten as

$$\nabla_{\frac{\partial}{\partial \bar z}}S_z = 0 \quad \text{or equivalently} \quad \nabla_{\frac{\partial}{\partial z}}S_{\bar z} = 0 \ . \tag{3.2}$$

The following is a basic existence and uniqueness theorem:

Theorem 3.1.1 (Eells, Sampson, 1964) *Suppose N is compact and has negative sectional curvature. Then $S \mapsto E_g(S)$ has a minimum in each homotopy class. Therefore, there exists a harmonic map in each homotopy class. Moreover, either this harmonic map is unique, or $S(M)$ is a closed geodesic in N. If $M = N$, we have uniqueness.*

Remark 3.1.1 *Uniqueness (due to Hartman [50] and Sampson [96]) follows from the negative curvature assumption. In the case $M = N$ the latter implies that the second variation at any critical point is positive, thus minima can be the only critical points. A mountain pass argument can be used to show that the assumption of more than one minimum leads to a contradiction by providing a saddle point between two minima. The original reference for theorem 3.1.1 is [28]. It is the existence part alone and needs 50 pages. A much shorter and more intuitive proof in the case $M = N$ can be found in the appendix.*

The following theorem makes Dirichlet's energy useful for our purposes; it is true in greater generality than is stated here (e.g. see [58]):

Theorem 3.1.2 (Schoen, Yau) *Suppose $N = M$ and $R(G) < 0$. Then the unique harmonic map $S : (M,g) \to (M,G)$ homotopic to the identity is an orientation preserving diffeomorphism, i.e.* $\begin{vmatrix} S_x^1 & S_y^1 \\ S_x^2 & S_y^2 \end{vmatrix} = (|S_z|^2 - |S_{\bar{z}}|^2) > 0$. *(A maximum principle is behind this last inequality.)*

Theorem 3.1.2 will be proved in the appendix, too.

Up to now we have considered Dirichlet's energy as a real valued function which is applied to a map on M, the metric g being a parameter. Fixing the map and stressing the dependence on g makes E a function defined on $\mathcal{M}$. Restricting it to $\mathcal{M}_{-1}$ and hoping for $\mathcal{D}_0$-invariance makes it a function on $T(M)$.

There are many Dirichlet's energies on $T(M)$ corresponding to the choice of (N, G). Here is the construction in detail:

Choose (N, G) with the sectional curvature of G negative and in such a way that the theorem of Eells and Sampson and uniqueness hold. Choose also a homotopy class of maps $M \to N$. Then the *unique* minimum $S(g)$ of $E_g(S)$ in this homotopy class will depend smoothly on g (see the following remark).

Define $\hat{E}(g) := E(g, S(g))$.

This is a smooth function on $\mathcal{M}$ or on $\mathcal{M}_{-1}$, and we claim it passes to a smooth function on $T(M) = \mathcal{M}_{-1}/\mathcal{D}_0$.

We have the following invariance property for classical Dirichlet's energy:

$$\text{For } f \in \mathcal{D}, \qquad E(f^*g, f^*S) = E(g, S) \ .$$

This follows from the fact that $f : (M, f^*g) \to (M, g)$ is holomorphic and Dirichlet's energy is invariant under holomorphic mappings.

We already know that $f^*g \in \mathcal{M}_{-1}$ if $g \in \mathcal{M}_{-1}$. Consider $\hat{E}(f^*g) = E(f^*g, S(f^*g))$. To continue we have to prove $S(f^*g) = f^*S(g)$. Then $\hat{E}(f^*g) = \hat{E}(g)$ follows immediately. This can be done for $f \in \mathcal{D}_0$ (but not for $f \in \mathcal{D}$) as follows:

$f^*S(g) = S(g) \circ f$ is harmonic from (M, f^*g) to (M, g), and it is homotopic to $S(g)$ if $f \in \mathcal{D}_0$. Uniqueness of the harmonic map shows that it is equal to $S(f^*g)$.

Remark 3.1.2 *That the unique minimum $S(g)$ depends smoothly on g is not trivial. Even in finite dimensional calculus, a non-degeneracy condition is needed as the example $a \mapsto \min_x \left(\frac{x^4}{4} + ax \right)$ shows. The classical concept of non-degeneracy in infinite dimensions, that the Hessian at the critical point is an isomorphism of the tangent space to its dual does not apply in our present context for the following reason: formally the Hessian should be the bilinear pairing $\int_M \nabla f \cdot \nabla g$, which could only be an isomorphism in the Sobolev space H^1. But a manifold $H^1(M, N)$ of maps from M to N does not exist in any intrinsic manner. We therefore cannot define intrinsically what it means for an H^1 function to have values in N.*
A weaker concept of non-degeneracy can be introduced [111] with the effect that, understood properly, the minimum is non-degenerate and therefore depends smoothly on g.
We can circumvent this difficulty here: our proof of the Eells-Sampson theorem will yield existence, uniqueness and smooth dependence.

Among all Dirichlet's energies on $\mathcal{T}(M)$ one could conceive of, we distinguish one and call it *the* Dirichlet energy.

It is defined by the above procedure with the choice $(N, G) = (M, g_0)$ where $g_0 \in \mathcal{M}_{-1}$. g_0 will be fixed throughout the exposition. That the choice of g_0 is arbitrary will play no role. Let $S(g)$ be the unique harmonic diffeomorphism from (M, g) to (M, g_0) which is homotopic to the identity.

We get $g \mapsto \tilde{E}(g) := E(g, S(g))$ as a function on $\mathcal{M}_{-1}$. Since $\tilde{E}(f^*g) = \tilde{E}(g)$ for $f^* \in \mathcal{D}_0$, $\tilde{E}$ can be considered as a function on $\mathcal{T}(M)$. We shall use the same letter $\tilde{E}$ for both functions.

We intend to show that Dirichlet's energy is a proper function and has a unique critical point. By an elementary Morse theoretic argument this guarantees that $\mathcal{T}(M)$ is a cell (i.e. diffeomorphic to some $\mathbb{R}^k$).

To this end, let us introduce our first theorem on Dirichlet's energy.

Theorem 3.1.3 *For $h \in T_{[g]}\mathcal{T}(M)$ let $\bar{h}$ be the horizontal lift of h to g. Let*

$$\xi(z) := \left\langle \frac{\partial S}{\partial x}, \frac{\partial S}{\partial x} \right\rangle - \left\langle \frac{\partial S}{\partial y}, \frac{\partial S}{\partial y} \right\rangle - 2i \left\langle \frac{\partial S}{\partial x}, \frac{\partial S}{\partial y} \right\rangle = \left\langle \frac{\partial S}{\partial z}, \frac{\partial S}{\partial z} \right\rangle = 4\rho S_z \overline{S_{\bar{z}}}$$

where $\langle \cdot, \cdot \rangle$ is the scalar product in TM coming from the metric g_0, which is in conformal coordinates $g_0 = ((\rho \delta_{ij}))$.

Then

> (i) $D\tilde{E}[g]h = -\frac{1}{4} \langle \operatorname{Re} \xi(z)dz^2, h \rangle_{WP} = -\frac{1}{2} \langle\langle \operatorname{Re} \xi(z)dz^2, \bar{h} \rangle\rangle$.
> In other words, $\operatorname{Re}\xi(z)dz^2$ considered as a tangent vector to $T(M)$ is the gradient of $\tilde{E}$ (up to a factor) with respect to the WP-metric.

> (ii) $[g_0]$ is the only critical point of $\tilde{E}$.

> (iii) $D^2\tilde{E}[g_0](h,k) = \langle h, k \rangle_{WP}$.
> In particular, g_0 is a non-degenerate critical point.

Remark 3.1.3 $D^2\tilde{E}$ is defined intrinsically only at a critical point.

PROOF: First note that $D\tilde{E}(g)\bar{h} = D\tilde{E}[g]h$. (The context makes it clear that $\tilde{E}$ is considered here as a function on $\mathcal{M}_{-1}$ on the left hand side and a function on $T(M)$ on the right hand side.)

Let us show that $\xi(z)$ is a holomorphic function:

$$\frac{\partial}{\partial \bar{z}} \left(\left\langle \frac{\partial S}{\partial x}, \frac{\partial S}{\partial x} \right\rangle - \left\langle \frac{\partial S}{\partial y}, \frac{\partial S}{\partial y} \right\rangle - 2i \left\langle \frac{\partial S}{\partial x}, \frac{\partial S}{\partial y} \right\rangle \right) =$$
$$2 \left\langle \frac{D}{\partial x} \frac{\partial S}{\partial x}, \frac{\partial S}{\partial x} \right\rangle - 2 \left\langle \frac{D}{\partial x} \frac{\partial S}{\partial y}, \frac{\partial S}{\partial y} \right\rangle + 2 \left\langle \frac{D}{\partial y} \frac{\partial S}{\partial x}, \frac{\partial S}{\partial y} \right\rangle + 2 \left\langle \frac{\partial S}{\partial x}, \frac{D}{\partial y} \frac{\partial S}{\partial y} \right\rangle + i(\ldots) \ .$$

The first and the fourth term cancel since S is harmonic, the second and the third term cancel since $\frac{D}{\partial x} \frac{\partial S}{\partial y} = \frac{D}{\partial y} \frac{\partial S}{\partial x}$. The calculation for the imaginary part goes similarly. So, the result is 0 as claimed. To fill in one more detail for the convenience of the reader, here is how the covariant derivatives entered the calculation: $\frac{DS}{\partial x} = \Pi(S(z)) \frac{\partial S}{\partial x}$, therefore

$$\left\langle \frac{\partial^2 S}{\partial x^2}, \frac{\partial S}{\partial x} \right\rangle = \left\langle \frac{\partial^2 S}{\partial x^2}, \Pi(S(z)) \frac{\partial S}{\partial x} \right\rangle = \left\langle \Pi(S(z)) \frac{\partial^2 S}{\partial x^2}, \frac{\partial S}{\partial x} \right\rangle = \left\langle \frac{D}{\partial x} \frac{\partial S}{\partial x}, \frac{\partial S}{\partial x} \right\rangle \ .$$

Here is another way to prove the same thing when conformal coordinates in the domain and range are used:

$$\begin{aligned}
\partial_{\bar{z}}(\rho S_z \overline{S_z}) &= (\rho_S S_z + \rho_{\bar{S}} \overline{S_z}) S_z \overline{S_z} + \rho S_{z\bar{z}} \overline{S_z} + \rho S_z \overline{S_{zz}} \\
&= (\rho_S S_z S_z + \rho S_{zz}) \overline{S_z} + (\rho_{\bar{S}} \overline{S_z} S_z + \rho \overline{S_{zz}}) S_z \\
&= 0
\end{aligned}$$

where the last line uses the equation (3.1) and ρ real ($\rho_S = \overline{\rho_S}$).

We shall need to use the derivative of S. So let $W(\check{h}) := DS(g)\check{h}$. (Here, W stands for "whatever".)

We have seen earlier that the derivative of $g \mapsto g^{ij}$ is $h \mapsto -h^{ij}$. An easy calculation shows that the derivative of $g \mapsto d\mu_g$ is $h \mapsto \left(\frac{1}{2}tr_g h\right) d\mu_g$. This allows us to calculate

$$D\tilde{E}(g)\check{h} = -\frac{1}{2}\int_M \check{h}^{ij}\frac{\partial S^\alpha}{\partial x^i}\frac{\partial S^\alpha}{\partial x^j}d\mu_g + \frac{1}{4}\int_M g^{ij}\frac{\partial S^\alpha}{\partial x^i}\frac{\partial S^\alpha}{\partial x^j}tr_g\check{h}\,d\mu_g + \int_M g^{ij}\frac{\partial W^\alpha}{\partial x^i}\frac{\partial S^\alpha}{\partial x^j}d\mu_g \quad . \quad (3.3)$$

The last term is nothing else but $DE_g(S)W = 0$ since S is a critical point of E_g. The second term vanishes because $\check{h}$ is horizontal, thus $tr_g\check{h} = 0$. For the first term note that in conformal coordinates on M, $g = ((\lambda\delta_{ij}))$, $\check{h}^{11} = -\check{h}^{22}, \check{h}^{12} = \check{h}^{21}$. Therefore

$$D\tilde{E}(g)\check{h} = -\frac{1}{2}\int_M \left\{\check{h}^{11}\left(\left\langle\frac{\partial S}{\partial x},\frac{\partial S}{\partial x}\right\rangle - \left\langle\frac{\partial S}{\partial y},\frac{\partial S}{\partial y}\right\rangle\right) + 2\check{h}^{12}\left\langle\frac{\partial S}{\partial x}\frac{\partial S}{\partial y}\right\rangle\right\} d\mu_g \quad .$$

On the other hand,

$$\xi(z)dz^2 = \left(\left\langle\frac{\partial S}{\partial x},\frac{\partial S}{\partial x}\right\rangle - \left\langle\frac{\partial S}{\partial y},\frac{\partial S}{\partial y}\right\rangle\right) dx^2 + 4\left\langle\frac{\partial S}{\partial x},\frac{\partial S}{\partial y}\right\rangle dx\,dy$$
$$- \left(\left\langle\frac{\partial S}{\partial x},\frac{\partial S}{\partial x}\right\rangle - \left\langle\frac{\partial S}{\partial y},\frac{\partial S}{\partial y}\right\rangle\right) dy^2 + i(\dots) \quad .$$

Thus $\text{Re}\,\xi(z)dz^2 = \rho_{11}dx^2 + 2\rho_{12}dx\,dy + \rho_{22}dy^2$ with ρ_{ij} defined by this equation. With this notation,

$$D\tilde{E}(g)\check{h} = -\frac{1}{2}\int_M \left(\check{h}^{11}\rho_{11} + \check{h}^{12}\rho_{12}\right)\lambda dx\,dy \quad .$$

But since $\check{h}^{11} = -\check{h}^{22}, \rho_{11} = -\rho_{22}$, we have

$$\langle h,\rho\rangle_{WP} = 2\int \left(\check{h}^{11}\rho_{11} + \check{h}^{12}\rho_{12}\right) d\mu_g \quad .$$

This proves (i).

Let us show next that $[g_0]$ is the only critical point of $\tilde{E}$. According to the formula just proved, $[g]$ critical implies $\text{Re}\,(\xi(z)dz^2) \equiv 0$ thus $\xi(z)dz^2 \equiv 0$ since ξ is holomorphic. We now really use that the range of S is two dimensional and introduce conformal coordinates on N:

$$g_0 = ((\rho\delta_{ij})), \quad \frac{1}{4}\xi(z)dz^2 = \rho S_z\overline{S_{\bar{z}}}dz^2 = \rho S_z(\bar{S})_z dz^2 \quad .$$

Since $\rho \neq 0$ everywhere, we get $S_z\overline{S_{\bar{z}}} = 0$. Together with $|S_z|^2 - |S_{\bar{z}}|^2 > 0$ which we get from theorem 3.1.2, it follows that $S_{\bar{z}} = 0$, thus S is holomorphic. Using Schoen-Yau again, S is a holomorphic *diffeomorphism* from (M,g) to (M,g_0), hence $c(g) = c(g_0)$ or in other words $[g] = [g_0]$. This proves (ii).

Even though theorem 3.1.2 is among the basic results we rely on repeatedly, it can be circumvented in the last argument. Let's argue that S is a holomorphic diffeomorphism from $S_z\overline{S_{\bar z}} = 0$, $\rho S_{z\bar z} + \rho_S S_z S_{\bar z} = 0$ and $S \sim id$ without using theorem 3.1.2. Being homotopic to the identity, S has degree 1 (and is therefore surjective); therefore its Jacobian determinant $|S_z|^2 - |S_{\bar z}|^2$ must be positive somewhere. (The degree is the sum of the signs of the Jacobian determinants at the pre-images of any given regular value, and regular values do exist due to Sard's theorem.) Then the determinant is positive in some neighbourhood, and one concludes that $S_{\bar z} = 0$ in this neighbourhood. The set K, where $S_{\bar z} = 0$, is therefore closed with non-empty interior. We claim it is dense as well (and hence all of M). Suppose not, then there is an open set $U = M\backslash K$ where $S_{\bar z} \neq 0$ and hence $S_z = 0$. Let B be the boundary of U and consider a neighbourhood V of a point in B. The harmonicity equation $S_{z\bar z} = 0$ holds in V. It is equivalent to $(\partial_x^2 + \partial_y^2)S^1 = (\partial_x^2 + \partial_y^2)S^2 = 0$. Hence S_z^1 and S_z^2 are holomorphic in V. On B, one has $S_z = S_{\bar z} = 0$, which is equivalent to $S_x^1 = S_y^1 = S_x^2 = S_y^2 = 0$. Hence S_z^1 and S_z^2 vanish on B. But B clearly has accumulation points in V, so, being holomorphic, S_z^1 and S_z^2 vanish in V. Therefore, so do $S_{\bar z}^1$ and $S_{\bar z}^2$ because S^1, S^2 are real valued. Thus $S_{\bar z} = 0$ in $U \cap V$, a non-empty (open) subset of U, contrary to the definition of U. Therefore, we have shown that S is holomorphic on all of M and hence it is either a diffeomorphism or it has branch points ($S_z = 0$). But the presence of branch points would imply a degree greater than 1. This last argument is known in greater generality as Hurwitz' formula in the theory of Riemann surfaces, namely $\chi(M) + r = \deg S \cdot \chi(M)$ where $\chi(M)$ is the Euler characteristic and r is the total order of branching. See e.g. [37, p. 128].

For (iii), when calculating the second derivative at g we should be aware that it is intrinsically defined as a bilinear map on $T_g\mathcal{M}$ only at a critical point g. Without such an assumption, the second derivative is a function on $T(T\mathcal{M})$, and the difficulty of defining it on $T_g\mathcal{M}$ stems from the fact that there is no intrinsic way to compare neighbouring tangent spaces. A pedestrian way to understand this fact is by considering the local situation in the image U of some coordinate chart mapping; U is open in some linear space. Let $f : U \to \mathbb{R}$, but think of f as defined on $\mathcal{M}$ via the coordinate map. Then the derivatives of f are given by the chain rule. For example, let $\varphi :]-\varepsilon, \varepsilon[\to M$, $\varphi(0) = x$, $\dot\varphi(0) = v$. Then Df could be defined by the equation $\frac{d}{dt}\big|_{t=0} f(\varphi(t)) = Df(\varphi(0))\dot\varphi(0)$. Similarly

$$\left(\tfrac{d}{dt}\right)^2\Big|_{t=0} f(\varphi(t)) = D^2 f(\varphi(0))(\dot\varphi(0), \dot\varphi(0)) + Df(\varphi(0))\ddot\varphi(0)$$

which is intrinsic if and only if $Df(\varphi(0)) = 0$ since $\dot\varphi(0)$ are tangent vectors, but in general $\ddot\varphi(0)$ are not.

We can cope with this formal difficulty here in quite a practical manner by extending $\tilde E$ from the manifold $\mathcal{M}_{-1}$ to a function $\hat E$ on the open set $\mathcal{M}$. The harmonic map $S(g)$

exists, is unique and depends smoothly on g for $g \in \mathcal{M}$ since the Eells-Sampson theorem needs the curvature assumption only in the range. This automatically defines us the extension $\hat{E}(g) := E(g, S(g)), g \in \mathcal{M}$. Since the tangent spaces to $\mathcal{M}$ and $\mathcal{M}_{-1}$ are different, we have to redo the calculation of the first derivative in a possibly non-horizontal direction h. (Being horizontal depends on g, which will vary for the calculation of the second derivative.)

The formula is the same as in (3.3). The second term no longer vanishes, but the third term still does. S is a critical point of E_g for all g, and the third term in (3.3) is the *partial* derivative of E_g with respect to S. Again using conformal coordinates in the domain, $g = ((\lambda \delta_{ij}))$, yields:

$$
\begin{aligned}
D\hat{E}(g)h &= -\frac{1}{2}\int h^{ij}\frac{\partial S^l}{\partial x^i}\frac{\partial S^l}{\partial x^j}d\mu_g + \frac{1}{2}\int g^{ij}\frac{\partial S^l}{\partial x^i}\frac{\partial S^l}{\partial x^j}\left(\frac{1}{2}tr_g h\right)d\mu_g \qquad (3.4)\\
&= -\frac{1}{2}\int \frac{1}{\lambda}H^i_j\frac{\partial S^l}{\partial x^i}\frac{\partial S^l}{\partial x^j}d\mu_g + \frac{1}{2}\int \frac{\delta^{ij}}{\lambda}\frac{\partial S^l}{\partial x^i}\frac{\partial S^l}{\partial x^j}\left(\frac{1}{2}tr H\right)d\mu_g\\
&= -\frac{1}{2}\int \lambda\left(H - \frac{1}{2}(trH)I\right)^i_j\frac{\partial S^l}{\lambda\partial x^i}\frac{\partial S^l}{\lambda\partial x^j}d\mu_g\\
&= -\frac{1}{2}\int g(x)(H^T\nabla S^l, \nabla S^l)d\mu_g .
\end{aligned}
$$

Here, the superscript T denotes the trace free part.

We have just proved the

Lemma 3.1.4
$$
\begin{aligned}
D\hat{E}(g)h &= -\frac{1}{2}\int g(x)(H^T\nabla_g S^l, \nabla_g S^l)d\mu_g\\
&= -\frac{1}{4}\langle\langle \operatorname{Re}\xi(z)dz^2, h^T\rangle\rangle .
\end{aligned}
$$

It implies that g_0 is critical for $\hat{E}$ as well as for $\bar{E}$. We want to calculate the second derivative $D^2\hat{E}(g)(h, k)$ for trace free h and k using (3.4). We need the derivative of $g \mapsto -h^{ij} = -g^{il}g^{jm}h_{lm}$. It is

$$
\begin{aligned}
k \mapsto k^{il}g^{jm}h_{lm} + g^{il}k^{jm}h_{lm} &= \frac{1}{\lambda}k^{il}h_{lj} + \frac{1}{\lambda}k^{jm}h_{im} = \frac{1}{\lambda}(k^l_i h^j_l + h^m_i k^j_m)\\
&= \frac{1}{\lambda}(KH + HK)\\
&= \frac{1}{\lambda}(h \cdot k)\cdot I .
\end{aligned}
$$

The last equality sign uses that h, k are 2-dimensional: If $K = \begin{pmatrix} a & b \\ b & -a \end{pmatrix}$, $H = \begin{pmatrix} c & d \\ d & -c \end{pmatrix}$, then $KH + HK = 2(ac + bd)I$. Next we need the derivative of $g \mapsto tr_g h = g^{ij}h_{ij}$. It is clearly $k \mapsto -k^{ij}h_{ij} = -h \cdot k$. We therefore get

$$
\begin{aligned}
D^2\hat{E}(g)(h,k) &= \tfrac{1}{2}\int_M \tfrac{1}{\lambda}(h \cdot k)\tfrac{\partial S^l}{\partial x^i}\tfrac{\partial S^l}{\partial x^i}d\mu_g - \int_M h^{ij}\tfrac{\partial W^i(k)}{\partial x^i}\tfrac{\partial S^l}{\partial x^i}d\mu_g \\
&\quad -\tfrac{1}{4}\int_M (h \cdot k)g(x)(\nabla_g S^l, \nabla_g S^l)d\mu_g \\
&= \tfrac{1}{4}\int_M (h \cdot k)g(x)(\nabla_g S^l, \nabla_g S^l)d\mu_g - \int_M h^{ij}\tfrac{\partial W^i(k)}{\partial x^i}\tfrac{\partial S^l}{\partial x^j}d\mu_g \ .
\end{aligned}
\tag{3.5}
$$

The second term vanishes at the critical point $g = g_0$ since according to the following lemma W vanishes.

Lemma 3.1.5 $W(h) = DS(g_0)h = 0$ for all h. *(This lemma does not hold necessarily for general Dirichlet energies at their critical points.)*

If $g = g_0$, then $S = id$ as the reader can easily verify. In this case, $g(x)(\nabla_g S^l, \nabla_g S^l) = 2$, and the first term simplifies to $D^2\hat{E}(g)(h,k) = \tfrac{1}{2}\langle\langle h,k\rangle\rangle_g$. Accepting the lemma for a minute, we can conclude the proof for the theorem: Let $\sigma(t)$ be any path in $\mathcal{T}(M)$ with $\sigma(0) = [g_0]$, $\sigma'(0) = h$. Call the horizontal lift of σ to $g_0 \in \mathcal{M}_{-1}$ again σ, so $\sigma(0) = g_0$, $\sigma'(0) = \bar{h}$ and use $\bar{E}(\sigma(t)) = \hat{E}(\sigma(t))$.

$$
\left(\frac{d}{dt}\right)^2 \hat{E}(\sigma(t)) = D^2\hat{E}(\sigma(t))(\sigma'(t),\sigma'(t)) + D\hat{E}(\sigma(t))\sigma''(t) \ .
$$

For $t = 0$, the second term vanishes by lemma 3.1.4 since at g_0, $\operatorname{Re}\xi(z)dz^2 \equiv 0$, and we get

$$
D^2\bar{E}[g](h,h) := \frac{d^2}{dt^2}\Big|_{t=0}\bar{E}(\sigma(t)) = \frac{1}{2}\langle\langle \bar{h},\bar{h}\rangle\rangle
$$

as claimed. (The formula with h,k follows from polarization.) ∎

PROOF OF LEMMA 3.1.5:   Repeat the equation for $S(g)$:

$$
\frac{1}{\sqrt{g}}\frac{\partial}{\partial x^i}g^{ij}\sqrt{g}\frac{\partial S^\alpha}{\partial x^j} + \Gamma^\alpha_{\gamma\beta}(S)\frac{\partial S^\gamma}{\partial x^i}\frac{\partial S^\beta}{\partial x^j}g^{ij} = 0 \ .
\tag{3.6}
$$

In principle, we have to take the total derivative of this equation with respect to g in a trace free direction h at $g = g_0$. Since this becomes horrible, we shall make use of the fact that the partial derivative with respect to S is closely connected with the second variation of $S \mapsto E_g(S)$, which is well known and need not be redone. The partial derivative with respect to g is easy since the terms $\sqrt{g}$ do not contribute anything for h trace free. The derivative of (3.6) with respect to g (ignoring the terms $S(g)$) is therefore (let $g = ((\lambda \delta_{ij}))$ and evaluate at $S = id$):

$$-\frac{1}{\sqrt{g}}\frac{\partial}{\partial x^i}h^{ij}\sqrt{g}\frac{\partial S^\alpha}{\partial x^j} - \Gamma^\alpha_{\gamma\beta}(S)\frac{\partial S^\gamma}{\partial x^i}\frac{\partial S^\beta}{\partial x^j}h^{ij} =$$

$$= -\frac{1}{\lambda}\frac{\partial}{\partial x^i}(\lambda h^{i\alpha}) - \Gamma^\alpha_{ij}(id)h^{ij} = -\frac{1}{\lambda}\frac{\partial}{\partial x^i}\left(\frac{1}{\lambda}h_{\alpha i}\right) - \frac{\Gamma^\alpha_{ij}(id)h_{ij}}{\lambda^2}\ .$$

Assuming that h is also divergence free (hence $\frac{\partial}{\partial x^i}(h_{\alpha i}/\lambda) = \frac{\partial(1/\lambda)}{\partial x^i}h_{\alpha i})$) and using that

$$\Gamma^\alpha_{ij}(id) = \frac{1}{2\lambda}\left(\frac{\partial\lambda}{\partial x^j}\delta_{i\alpha} + \frac{\partial\lambda}{\partial x^i}\delta_{j\alpha} - \frac{\partial\lambda}{\partial x^\alpha}\delta_{ij}\right)\ ,$$

all terms cancel.

Denoting the left hand side of (3.6) by $L(g, S(g))$ for the moment, we have seen that

$$0 = \frac{d}{dg}L\left(g, S(g)\right) = \underbrace{\frac{\partial}{\partial g}L(g, S)}_{0} + \frac{\partial}{\partial S}L(g, S)\cdot DS(g)\ .$$

Let $g = g_0$, $S = id$, $DS(g_0)h = W$ and take the g_0-inner product of the equation with $W(h)$ and integrate to get

$$0 = \int\frac{\partial L}{\partial S}(g_0, id)W\cdot W = D^2E_g(id)(W, W)$$

Here is the formula for the second variation of E_g according to [76]:

$$D^2E_g(id)(W, W) = \int\left(\left\|\nabla_{\frac{\partial}{\partial x}}W\right\|^2 + \left\|\nabla_{\frac{\partial}{\partial y}}W\right\|^2\right)dx\,dy$$
$$-\int g_0\left(\mathcal{R}\left(\frac{\partial S}{\partial x}, W\right)W, \frac{\partial S}{\partial x}\right)dx\,dy \qquad (3.7)$$
$$-\int g_0\left(\mathcal{R}\left(\frac{\partial S}{\partial y}, W\right)W, \frac{\partial S}{\partial y}\right)dx\,dy\ ,$$

where $\mathcal{R}$ denotes the curvature operator for (M, g_0). It appears as a sectional curvature in the formula. These sectional curvatures are ≤ 0 because we are in two dimensions and the scalar curvature is negative. Therefore $D^2E_g(id)(W, W) = 0$ implies that all terms in the sum vanish separately, hence (since $S = id$) $W = 0$. ∎

Remark 3.1.4 *For an elegant and quick proof of lemma 3.1.5 see remark 6.2.4.*

3.2   The Properness of Dirichlet's Energy

Next we intend to show that $\tilde{E} : T(M) \to \mathbb{R}$ is a proper function, i.e. the inverse image of compact sets is compact. A Morse theory type argument immediately allows one to conlude from this that M is a cell. We shall use two basic lemmas whose proof is deferred to the appendix. Our first lemma is from hyperbolic geometry. Let $\mathbb{H}$ be the upper half plane equipped with the metric $ds^2 = 2\frac{dx^2+dy^2}{y^2}$ giving the constant scalar curvature -1 (i.e. Gauss curvature $-1/2$). This is a model for the hyperbolic plane, the universal cover of the manifolds M we consider. We use euclidean polar coordinates (r,θ) on $\mathbb{H}$: $x = r\cos\theta$, $y = r\sin\theta$.

Lemma 3.2.1 (Collar Lemma) *Let γ be a (non-trivial) closed geodesic of length ℓ on a surface (M,g) where $g \in \mathcal{M}_{-1}$. Then there exists a neighbourhood U of γ in M which is isometric to the following set $T/\sim$ in the hyperbolic plane:*

$$T = \left\{ (r,\theta) \mid 1 \le e \le r^{\ell/\sqrt{2}}, \theta_0 \le \theta \le \pi - \theta_0 \right\}$$

and $\sim$ identifying $(1,\theta)$ with $(e^{\ell/\sqrt{2}},\theta)$. Here, ℓ and θ_0 satisfy the estimate

$$\cot^4 \frac{\theta_0}{2} \ge \frac{1 + \cosh(\ell/\sqrt{2})}{\sinh(\ell/\sqrt{2})} \ .$$

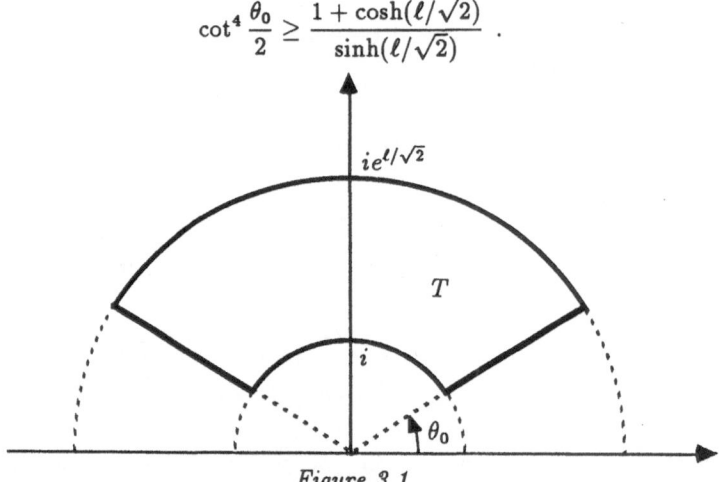

Figure 3.1

The point of this lemma is the following: the curvature condition guarantees that the shorter a geodesic is, the wider the collar is that can be placed around it.

We present the following lemma in two versions. They are obviously closely related though a proof would be required to show that both versions are equivalent. We shall not give the proof of the first (original) version. The second version is the one we need and we give its proof in the appendix.

Lemma 3.2.2 (Mumford Compactness Theorem)

FIRST VERSION: *Let* $(M, g_n) = \mathbb{H}/\Gamma_n$ *where* $\Gamma_n \subset SL_2(\mathbb{R}) = \left\{ z \mapsto \frac{az+b}{cz+d} \big| ad - bc = 1 \right\}$. *Suppose that all non-trivial closed geodesics on* (M, g_n) *are of length uniformly bounded below away from 0. Then there exists a subsequence* (Γ_m) *of* (Γ_n) *and a* $\Gamma \subset SL_2(\mathbb{R})$ *such that* $\Gamma_m \to \Gamma$ *and* $\mathbb{H}/\Gamma$ *is a Riemann surface of the topological type of* M. *See [79].*

SECOND VERSION: *Let* (M, g_n), $g_n \in \mathcal{M}_{-1}$ *be a sequence of Riemann surfaces such that all non-trivial closed geodesics on* (M, g_n) *are of length uniformly bounded away from 0. Then there exists a subsequence* (g_m) *of* (g_n) *and* $f_m \in \mathcal{D}$ *and* $g \in \mathcal{M}_{-1}$ *such that* $f_m^* g_m \xrightarrow{C^\infty} g$.

Remark 3.2.1 *The lemma is not true if* $\mathcal{D}$ *is replaced by* $\mathcal{D}_0$. *We can immediately see the reason:* $\mathcal{D}/\mathcal{D}_0$ *is an infinite discrete group (hence non-compact). So if we choose* $h_n \in \mathcal{D}$ *such that* $[h_n] \in \mathcal{D}/\mathcal{D}_0$ *are all distinct and have no convergent subsequence and let* $g_n := h_n^* g_0$, *the boundedness condition is certainly satisfied for* g_n *since all* (M, g_n) *are isometric to* (M, g), *the isometry being* h_n. *But if we could find* $f_n \in \mathcal{D}_0$ *such that* $f_m^* h_m^* g_0 \to g$, *properness of the action would imply* $h_m \circ f_m \to : f$ *for a further subsequence. But* $[h_m \circ f_m] = [h_m]$, *so convergence contradicts our assumption.*

According to the remark, the Mumford compactness theorem is appropriate for the Riemann moduli space but apparently not for the Teichmüller moduli space. We shall later see however, how this gap can be bridged.

Keeping in mind the Arzelà-Ascoli theorem, the reader will no doubt believe that the following lemma can be helpful in proving that $\bar{E}$ is proper:

Lemma 3.2.3 *Let* $g \in \mathcal{M}_{-1}$ *be fixed and let* $v_n : M \to M$ *be a sequence of diffeomorphisms such that* $E(g, v_n) \leq K$. *Then the sequence* (v_n) *is equicontinuous.*

PROOF: Before coming to the main argument, let us fix the notation: if (x, y) are conformal coordinates, then Dirichlet's energy is $E_g(v) = \frac{1}{2} \int \left(\left\| \frac{\partial v}{\partial x} \right\|^2 + \left\| \frac{\partial v}{\partial y} \right\|^2 \right) dx \, dy$, where $\|\cdot\|$ represents the norm on the ambient Euclidian space in which (M, g_0) is embedded. We are going to use polar coordinates (r, θ) with respect to conformal coordinates (x, y), i.e.

$x = r\cos\theta$, $y = r\sin\theta$. These coordinates are therefore *not geodesic* polar coordinates. They have the effect instead that all formulas look as if M were flat. A ball $B_\delta(z_0)$ will denote the set where $r \leq \delta$ (defined with respect to a coordinate chart). Let $\delta < 1$ and small enough such that all balls $B_{\sqrt{\delta}}(z_0) \subset M$ are contained in a coordinate chart.

Claim: There exists $\rho \in [\delta, \sqrt{\delta}]$ such that

$$\int_0^{2\pi} \left\|\frac{\partial v_n(\rho,\theta)}{\partial\theta}\right\| d\theta \leq \left(\frac{8\pi K}{\log\frac{1}{\delta}}\right)^{1/2} \leq o(\delta) \ .$$

This claim is often referred to as the Courant-Lebesgue lemma.

To prove the claim suppose $\int_0^{2\pi} \left\|\frac{\partial v_n(\rho,\theta)}{\partial\theta}\right\| d\theta \geq m^{1/2}$ for all ρ. We have to show that this implies $m \leq \frac{8\pi K}{\log\frac{1}{\delta}}$. Then

$$
\begin{aligned}
\left(\int_\delta^{\sqrt{\delta}} \frac{dr}{r}\right) \cdot m \ &\leq\ \int_\delta^{\sqrt{\delta}} \frac{1}{r}\left(\int_0^{2\pi}\left\|\frac{\partial v_n(\rho,\theta)}{\partial\theta}\right\|\right)^2 d\theta\, dr \\
&\leq\ 2\pi \int_0^{2\pi}\int_\delta^{\sqrt{\delta}} \frac{1}{r^2}\left\|\frac{\partial v_n(\rho,\theta)}{\partial\theta}\right\|^2 r\, dr\, d\theta \\
&\leq\ 4\pi \int_{B_{\sqrt{\delta}}(z_0)\setminus B_\delta(z_0)} \frac{1}{2}\left(\frac{1}{r^2}\left\|\frac{\partial v_n}{\partial\theta}\right\|^2 + \left\|\frac{\partial v_n}{\partial r}\right\|^2\right) r\, dr\, d\theta \\
&\leq\ 4\pi E(g, v_n) \ .
\end{aligned}
$$

On the other hand $\int_\delta^{\sqrt{\delta}} \frac{dr}{r} = \frac{1}{2}\log\frac{1}{\delta}$. This proves the Courant-Lebesgue lemma.

The Courant-Lebesgue lemma now implies equicontinuity as follows: suppose $x, y \in M$ are δ-close. Then there exists a ball $B_\delta(z_0)$ containing them both. Then choose ρ according to the above formula. The points $v_n(x), v_n(y)$ both lie in $v_n\left(B_\rho(z_0)\right)$, which is a set diffeomorphic to a disc whose perimeter is bounded by $o(\delta)$ independent of v_n, z_0. Their distance is then bounded by the same constant. ∎

We are now well prepared to attack

Theorem 3.2.4 $\bar{E} : T(M) \to \mathbb{R}$ *is proper.*

PROOF: Suppose $\bar{E}[g_n] = E\left(g_n, S(g_n)\right)$ tends to some limit, hence $E\left(g_n, S(g_n)\right) \leq K$. We must show that $([g_n])$ has a convergent subsequence. Abbreviate $S(g_n) =: S_n$. Since all S_n are homotopic to the identity, they map homotopically non-trivial loops γ into homotopically non-trivial loops $S_n(\gamma)$. The latter are all in (M, g_0) with g_0 one and the same for all S_n, hence their lengths are trivially bounded from below by some constant δ independent of n, γ. We claim that this fact, together with the bound for Dirichlet's energy also establishes a lower bound for the length ℓ of the loop γ, namely that

$$\ell \geq \frac{\delta^2(\pi - 2\theta_0)}{\sqrt{2}K} \; , \tag{3.8}$$

where θ_0 corresponds to ℓ as in the collar theorem. Postponing the proof of this claim for a moment, we can apply Mumford's compactness theorem to find a subsequence $f_m \in \mathcal{D}$ such that $f_m^* g_m =: \hat{g}_m$ converges to, say, g. We are not yet finished because presumably $f_m \notin \mathcal{D}_0$ so we cannot conclude that $[g_m]$ converges in $\mathcal{T}(M)$. What we *can* conclude is that the larger equivalence classes of g_m in the *Riemann* moduli space converge.

What helps at this point is the observation that the sequence (f_m) runs through only finitely many equivalence classes in $\mathcal{D}/\mathcal{D}_0$. This will be shown in a moment. Assuming this, then by selecting a further subsequence it is no loss of generality to assume that all f_m lie in the same equivalence class, i.e their orbit spaces are equal: $\mathcal{D}_0 f_m = \mathcal{D}_0 f$. Hence $f_m = h_m \circ f$ with $h_m \in \mathcal{D}_0$, and $h_m^* g_m = (f^{-1})^* f_m^* g_m \to (f^{-1})^* \hat{g}$, i.e. $[g_n] \to [(f^{-1})^* \hat{g}]$.

Why does the sequence (f_m) run through only finitely many equivalence classes in $\mathcal{D}/\mathcal{D}_0$? If not, a further subsequence (called (f_m) again) would run through pairwise different classes giving rise to a sequence of diffeomorphisms $v_m = S_m \circ f_m$ belonging to different homotopy classes, but nevertheless equicontinuous by lemma 3.2.3. This lemma applies because

$$K \geq E(g_m, S_m) = E(f_m^* g_m, f_m^* S_m) = E(f_m^* g_m, v_m)$$

and remembering that the very first assumption was $f_m^* g_m \to g$, we have for large m, $E(g, v_m) \leq 2K$.

But a sequence v_m belonging to distinct homotopy classes cannot be equicontinuous, because some subsequence of an equicontinuous sequence converges (Arzelà-Ascoli), hence its homotopy class is eventually constant.

We are left with the proof of the estimate (3.8)

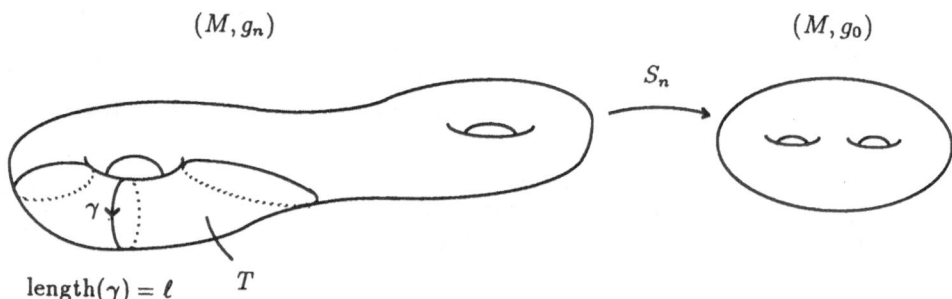

(M, g_n) (M, g_0)

S_n

γ

length$(\gamma) = \ell$ T

Figure 3.2:

To this end use the collar around γ provided by lemma 3.2.1. For all $\theta_0 \leq \theta \leq \pi - \theta_0$ it holds

$$\int\limits_{1}^{e^{\ell/\sqrt{2}}} \left\| \frac{\partial S_n}{\partial r}(r, \theta) \right\| dr \geq \delta$$

because this integral is the length of a non-trivial loop in (M, g_0).

Therefore, Cauchy-Schwarz implies

$$\delta^2 \leq \left(\int\limits_{1}^{e^{\ell/\sqrt{2}}} \left\| \sqrt{r} \frac{\partial S_n}{\partial r}(r, \theta) \right\| \cdot \frac{1}{\sqrt{r}} \, dr \right)^2 \leq \left(\int\limits_{1}^{e^{\ell/\sqrt{2}}} \frac{dr}{r} \right) \left(\int\limits_{1}^{e^{\ell/\sqrt{2}}} r \left\| \frac{\partial S_n}{\partial r}(r, \theta) \right\|^2 dr \right)$$

and, integrating this with respect to θ (i.e. over the collar),

$$\begin{aligned} \delta^2(\pi - 2\theta_0) &\leq \frac{1}{\sqrt{2}} \int\limits_{\theta_0}^{\pi-\theta_0} \int\limits_{1}^{e^{\ell/\sqrt{2}}} \left(\left\| \frac{\partial S_n}{\partial r} \right\|^2 + \frac{1}{r^2} \left\| \frac{\partial S_n}{\partial \theta} \right\|^2 \right) r \, dr \, d\theta \\ &\leq \frac{\ell}{\sqrt{2}} \cdot 2E(g_n, S_n) \\ &\leq \ell\sqrt{2}K \ . \end{aligned}$$

This immediately gives (3.1.5), thus also ending the proof of the theorem. ∎

3.3   Teichmüller Space is a Cell

The basic idea for our proof that Teichmüller space is a cell uses a simple result from differential topology, namely

Theorem 3.3.1 *Let $\tilde{E} : \mathcal{T} \to \mathbb{R}^+$ be a C^∞ smooth real valued functional on a finite m-dimensional manifold $\mathcal{T}$ satisfying the following properties:*

(i) *$\tilde{E}$ has only one critical point $p_0 \in \mathcal{T}$;*

(ii) *$\tilde{E}$ is a proper function; that is the inverse image of any closed interval $\tilde{E}^{-1}[a, b]$ is a compact set in $\mathcal{T}$.*

(iii) *p_0 is a non-degenerate critical point, that is*

$$D^2 \tilde{E}_{p_0} : T_{p_0}\mathcal{T} \times T_{p_0}\mathcal{T} \to \mathbb{R}$$

is a non-degenerate quadratic form.

Then $\mathcal{T}$ is diffeomorphic to $\mathbb{R}^m$.

PROOF: We can assume without loss of generality that $\tilde{E}(p_0) = 0$. Furthermore we can always assume that $\mathcal{T}$ has a Riemannian metric $\langle \cdot, \cdot \rangle_p : T_p\mathcal{T} \times T_p\mathcal{T} \to \mathbb{R}$. Let $\nabla \tilde{E}$ denote the gradient of $\tilde{E}$ with respect to this metric and for $p \neq p_0$, let $\sigma_p(t)$ be the flow of the vector field $Z = \frac{\nabla \tilde{E}}{\|\nabla \tilde{E}\|^2}$. Thus

$$\frac{d}{dt}\sigma_t(t) = Z\left(\sigma_p(t)\right) , \quad \sigma_p(0) = p . \tag{3.9}$$

Then

$$\frac{d}{dt}\tilde{E}\left(\sigma_p(t)\right) = \left\langle D\tilde{E}\left(\sigma_p(t)\right), \frac{d}{dt}\sigma_p(t) \right\rangle = 1 ,$$

whence we see that

$$\tilde{E}\left(\sigma_p(t)\right) = t + \tilde{E}(p) . \tag{3.10}$$

Since p_0 is the only critical point of $\tilde{E}$ it must be the absolute minimum (which exists by properness). Moreover p_0 is a non-degenerate critical point so by the famous Morse lemma [76] we may assume that there is a coordinate system $\varphi : U \to \mathcal{T}$ about p_0 in which $\tilde{E}$ has the local form

$$(\tilde{E} \circ \varphi)(x_1, \ldots, x_m) = \frac{1}{2}(x_1^2 + \ldots + x_m^2) , \tag{3.11}$$

therefore for small ε, $\tilde{E}^{-1}[0, \varepsilon[$ is diffeomorphic to an open ball B_ε. Pick $\varepsilon > 0$ so small that $B_{3\varepsilon} \subset$ range φ. Since, in general, Riemannian metrics are constructed from partitions of unity, we can assume that $\langle \cdot, \cdot \rangle$ restricted to $B_{2\varepsilon}$ is the pullback via φ of the Euclidean metric on $\mathbb{R}^m$.

By a partition of unity we can construct a vector field X as a convex combination of $\nabla \tilde{E}$ and Z so that

(i) $D\tilde{E}(p)(X(p)) > 0$ except for $p = p_0$ (where $X(p) = 0$); moreover, $D\tilde{E}(p)(X(p)) > \delta$ on $B_{3\epsilon} \setminus B_\epsilon$

(ii) $X \mid B_{2\epsilon} = \nabla\tilde{E}$

(iii) $X \mid \mathcal{T} \setminus B_{3\epsilon} = Z$.

Again let $\sigma_p(t)$ denote the flow of X. For $p \in \mathcal{T} \setminus B_{3\epsilon}$ it is easy to see that formula (3.10) holds for all positive time. For $p \in B_{2\epsilon}$, let $\varphi(p) = x$; then by formula (3.11) we have that locally $\nabla\tilde{E}$ can be expressed as

$$\nabla\tilde{E}(x_1, \ldots, x_m) = (x_1, \ldots, x_m) .$$

Therefore in this local coordinate neighbourhood $\sigma_p(t)$ can be expressed as

$$\sigma_p(t) = \varphi^{-1}(e^t x) . \tag{3.12}$$

For arbitrary p let $t^+(p)$ and $t^-(p)$ denote the maximal positive and negative time that the flow $\sigma_p(t)$ is defined. The following lemma is standard (e.g. [64]).

Lemma 3.3.2 Let $X : \mathcal{T} \to T\mathcal{T}$ be a C^1 vector field on a manifold $\mathcal{T}$ without boundary. If either $t^+(p) < \infty$ or $t^-(p) > -\infty$, then for any sequence $t_n \to t^+(p)$, or $t_n \to t^-(p)$, $\sigma_p(t_n)$ cannot have a limit point in $\mathcal{T}$.

Lemma 3.3.3 If $p \in \mathcal{T}$, $t^-(p) = -\infty$, $t^+(p) = +\infty$.

PROOF: If $t^+(p) < \infty$, then we know by formula (3.10), that $\{\sigma_p(t) \mid 0 \leq t \leq t^+(p)\}$ must be in $\tilde{E}^{-1}[0, t^+(p) + 1]$, a compact set. Thus for any sequence $t_n \to t^+(p)$, $\sigma_p(t_n)$ must have a limit point contradicting lemma 3.3.2. Similarly, we can argue that $t^-(p) = -\infty$.

Lemma 3.3.4 For all $p \in \mathcal{T}$, $\lim_{t \to \infty} \sigma_p(t) = p_0$.

PROOF: $\tilde{E}(\sigma_p(t))$ strictly decreases as $t \to -\infty$ according to formula (3.10) until eventually $\tilde{E}(\sigma_p(t)) \leq 3\epsilon$. Then condition (i) above guarantees that for some t_0 large enough, $\sigma_p(t_0) = q \in B_{2\epsilon}$. Then locally, $\sigma_s(q) = e^s y$, $y = \varphi(q)$. Clearly by formula (3.12), $\sigma_p(t_0 + s) = \sigma_s(q) \to p_0$ as $s \to -\infty$. ∎

We are now ready for the proof of our theorem. Let $\varphi : [0, \varepsilon[\rightarrow [0, \infty[$ be a C^∞ diffeomorphism with $\varphi(t) = t$ for $0 \leq t \leq \varepsilon/4$.

Define a diffeomorphism $\nu : B_{2\varepsilon} \rightarrow T$ by setting $\nu\big|_{B_\varepsilon} = id$. Then clearly any $q \in B_{2\varepsilon} \setminus B_\varepsilon$ can be written uniquely as $q = \sigma_p(t)$, $0 \leq t \leq \log 2$ (again see formula (3.12)).

Define $\nu(q) = \sigma_p(\varphi(t))$. ν is clearly C^∞ on $T \setminus \partial B_\varepsilon$ and further observation shows that ν is C^∞ on ∂B_ε as well. ∎

Applying the lemma to Dirichlet's energy $\bar{E}$ on $T(M)$ we see immediately that Teichmüller space is a cell.

3.4 Topological Implications; The Contractibility of $\mathcal{D}_0$

The principal bundle $(\pi, \mathcal{D}_0, \mathcal{M}_{-1}, T(M))$ is in fact a trivial bundle, i.e. it has global sections. A section is a map $\sigma : T(M) \rightarrow A$ such that $\pi \circ \sigma = id$. We shall consider two of them. The first is the Earle-Eells section [22]:

Let $(M, g_0), (M, g) \in \mathcal{M}_{-1}$. Let $s(g) : (M, g_0) \rightarrow (M, g)$ be the unique harmonic map homotopic to the identity. It depends smoothly on g, and is again a diffeomorphism (Appendix B). Note that in contrast to the previous sections of this chapter we consider here the metric on the range as varying and the one on the domain as fixed. We consider this situation again in section 6.3. Define $\sigma_{EE}(g) := s(g)^*g$. Since $\sigma_{EE}(f^*g) = \sigma_{EE}(g)$ for $f \in \mathcal{D}_0$ as we shall immediately see, σ_{EE} is indeed a map from $T(M)$ to $\mathcal{M}_{-1}$, and obviously $\pi \circ \sigma_{EE} = id$.

Since $f : (M, f^*g) \rightarrow (M, g)$ is an isometry and harmonic maps remain harmonic when composed from the left with isometries, $f \circ s(f^*g)$ is a harmonic map from (M, g_0) to (M, g), and it is homotopic to the identity, if f is. Hence $f \circ s(f^*g) = s(g)$ and

$$\sigma_{EE}(f^*g) = s(f^*g)^* f^*g = \left(f^{-1} \circ s(g)\right)^* f^*g = s(g)^*(f^{-1})^* f^*g = \sigma_{EE}(g)$$

as claimed.

A different section can be defined similarly by defining $S(g)$ as in section 3.1 to be the harmonic map from (M, g) to (M, g_0) homotopic to the identity and letting $\sigma(g) := S(g)_*g$. This definition again uses the fact that $S(g)$ is a diffeomorphism, for else the pushforward $S(g)_*$ would not be defined. In this case we have

$$\sigma(f^*g) = S(f^*g)_* f^*g = (S(g) \circ f)_* f^*g = S(g)_* f_* f^*g = \sigma(g) \ .$$

Another implication is that $\mathcal{D}_0$ is contractible. One could prove this by making use of the result from the theory of fibre bundles (π, F, E, B) that if any two of base space B, total space E and fibre F are contractible, then so is the third, provided the base is of the homotopy type of a path connected CW-complex. Such a result follows by first constructing from the fibration $F \hookrightarrow E \to B$ a fibration up to homotopy

$$\cdots \to \Omega^n F \to \Omega^n E \to \Omega^n B \to \Omega^{n-1} F \to \cdots$$

of the loop spaces according to theorem 4.20 of [71]. Applying the functor $[Z, \cdot]$ (homotopy classes of mappings from Z into a space) to that subsequence of length 3 of

$$\Omega E \to \Omega B \to F \to E \to B$$

for which we know that the spaces at both ends are contractible, and choosing Z to be the middle space, theorem 4.19 of [71] (a sort of exactness of the resulting sequence) guarantees the contractibility of the middle space. In case the middle space is ΩB we still have to conclude from its contractibility the fact that B is contractible. This is not generally true but holds for path connected CW-complexes:

$$\Omega B \text{ contractible} \quad \Rightarrow \quad \pi_n(\Omega B) = \pi_{n+1}(B) = 0 \text{ for all } n \quad \Rightarrow \quad B \text{ contractible.}$$

Using this reasoning we can argue that $\mathcal{M}$ and $\mathcal{P}$ are contractible since they are cones in vector spaces. Hence $\mathcal{M}_{-1} \cong \mathcal{M}/\mathcal{P}$ is contractible, too. Now $T(M)$ is a cell, hence contractible, and since $T(M) \cong \mathcal{M}_{-1}/\mathcal{D}_0$, $\mathcal{D}_0$ must be contractible, too.

We need however not use this machinery; here is an elementary argument: $\mathcal{M}$ is contractible, and the bundle $(\Xi, \mathcal{P}, \mathcal{M}, \mathcal{M}_{-1})$ has an obvious section, namely the inclusion $\iota : \mathcal{M}_{-1} \hookrightarrow \mathcal{M}$. Then if $F : [0,1] \times \mathcal{M} \to \mathcal{M}$ is a contraction (i.e. $F(1, \cdot) = const$ and $F(0, \cdot) = id$, then

$$F_{-1} : [0,1] \times \mathcal{M}_{-1} \to \mathcal{M}_{-1}, \quad (t, g) \mapsto \pi(F(t, \iota(g)))$$

is a contraction of $\mathcal{M}_{-1}$. Hence $\mathcal{M}_{-1}$ is contractible. $T(M)$ is also contractible, therefore every bundle over $T(M)$ is trivial; in our case, we have even explicitly constructed a section $\sigma : T(M) \to \mathcal{M}_{-1}$, and hence a diffeomorphism

$$T : \mathcal{D}_0 \times T(M) \to \mathcal{M}_{-1}, \quad (f, [g]) \mapsto f^*\sigma[g] \ .$$

It gives rise to a projection map $pr : \mathcal{M}_{-1} \to \mathcal{D}_0$. The contraction F_{-1} of $\mathcal{M}_{-1}$ gives for any choice of $[g_0] \in T(M)$ a contraction

$$F_D : [0,1] \times \mathcal{D}_0 \to \mathcal{D}_0, \quad (t, f) \mapsto pr\left(F_{-1}(t, T(f, [g_0]))\right)$$

of $\mathcal{D}_0$.

4 The Complex Structure on Teichmüller Space

In this chapter we are going to consider the following question: Does $T(M)$ have a natural almost complex structure? Let us pose this question differently: does $\mathcal{A}$ have a natural almost complex structure, and if so, does it pass to the quotient? We will answer these questions in the affirmative. Contrary to the case of two real dimensions, it is not true in higher dimensions that an almost complex structure induces a complex structure. Therefore the next question is whether $T(M)$ has a complex structure. It turns out, it has. In order that an almost complex structure yields a complex structure it is necessary and sufficient that a certain tensor, the Nijenhuis tensor, vanishes identically. The Nijenhuis tensor plays a similar role for this type of question as does the Riemann curvature tensor in Riemannian geometry.

We have already defined $\mathcal{A} = \{J \mid J^2 = -id\} \subset C^\infty(T_1^1 M)$, and already know that $T_J \mathcal{A} = \{H \mid HJ = -JH\} \subset C^\infty(T_1^1 M)$ where J and H are 1-1 tensors on M. It is natural to associate with each $J \in \mathcal{A}$ a map $\Phi_J : T_J \mathcal{A} \to T_J \mathcal{A}$ by $\Phi_J(H) = JH$. At every $J \in \mathcal{A}$, Φ_J is a linear map from the tangent space at that point into itself, i.e. a 1-1 tensor on $\mathcal{A}$; clearly $\Phi_J^2 = -id$, so Φ is an almost complex structure on $\mathcal{A}$. We will call it the natural almost complex structure on $\mathcal{A}$.

4.1 Almost Complex Principal Fibre Bundles

In order to further attack the question posed at the beginning of the chapter, we have to develop a little general theory:

Definition 4.1.1 N is called a complex manifold of complex dimension m, if it is a manifold and can be covered by coordinate charts (φ_i, U_i), $U_i \subset \mathbb{C}^m$ such that $\varphi_i \circ \varphi_j^{-1}$ are holomorphic (where defined).

Definition 4.1.2 N is called an almost complex manifold of real dimension $2m$, if for each $p \in N$ there exists $J_p : T_pN \to T_pN$ such that $J_p^2 = -id$.

Remark 4.1.1 *On an almost complex manifold, an orientation is automatically determined in the following manner: choose independent vector fields $X_1, \ldots, X_m$ in a neighbourhood of every point such that $(X_1, \ldots, X_m, JX_1, \ldots, JX_m)$ is a basis. This is possible by inductively choosing X_k linearly independent of $(X_1, \ldots, X_{k-1}, JX_1, \ldots, JX_{k-1})$. Then $\sum_{i=1}^{k}(a_iX_i + b_iJX_i) = 0$ implies through left multiplication by $-a_k + b_kJ$ together with the linear independence of $(X_1, \ldots, X_k, JX_1, \ldots, JX_{k-1})$ that $a_k^2 + b_k^2 = 0$, hence $(X_1, \ldots, X_k, JX_1, \ldots, JX_k)$ is also linearly independent.*
All bases of the form $(X_1, \ldots, X_m, JX_1, \ldots, JX_m)$ are compatibly oriented. To see this, suppose $Y_i = \sum_j(a_{ij}X_j + b_{ij}JX_j)$, hence $JY_i = \sum_j(-b_{ij}X_j + a_{ij}JX_j)$. Letting $A = ((a_{ij}))$, $B = ((b_{ij}))$ we have to show that the determinant of the block matrix $\begin{bmatrix} A & B \\ -B & A \end{bmatrix}$ is positive. Here is the calculation:

$$
\begin{aligned}
\begin{vmatrix} A & B \\ -B & A \end{vmatrix} &= \begin{vmatrix} A & B \\ 0 & A+BA^{-1}B \end{vmatrix} = \det A \det(A + BA^{-1}B) = \det(A^2 + ABA^{-1}B) = \\
&= \det\left((A - iABA^{-1})(A + iB)\right) = \det A \det(A - iB)\det A^{-1}\det(A + iB) \\
&= \left|\det(A - iB)\right|^2 \geq 0 .
\end{aligned}
$$

In the last step, we have used that A, B are real matrices. The calculation holds for regular A, but the result holds for all A by continuity.
The determinant is actually positive since the matrix describes a change of bases.

Similarly as in section 1.1 an almost complex structure J can be associated to a complex structure as follows: in local coordinates $z^1, \ldots, z^m$ (where $z^\alpha =: x^\alpha + iy^\alpha$) define

$$
J\frac{\partial}{\partial x^\alpha} = \frac{\partial}{\partial y^\alpha} \quad ; \quad J\frac{\partial}{\partial y^\alpha} = -\frac{\partial}{\partial x^\alpha} .
$$

Definition 4.1.3 Given an almost complex structure J on N, a 1-2 tensor $\mathbf{N}(J)$, called the Nijenhuis tensor, is associated to it as follows:

$$
\mathbf{N}(J) : \mathcal{X}(N) \times \mathcal{X}(N) \to \mathcal{X}(N)
$$
$$
\mathbf{N}(J)(X, Y) = 2([JX, JY] - [X, Y] - J[JX, Y] - J[X, JY])
$$

where $[X, Y]$ is the commutator of vector fields, and $\mathcal{X}(N)$ are the C^∞ vector fields on N.

Let us explicitly work out the definition of the commutator for the reader's convenience by presenting the general argument in such a way as it appears in the situation at hand. The manifold $\mathcal{A}$ is embedded in the vector space $E := C^\infty(T_1^1 M)$, so vector fields X, Y on $\mathcal{A}$ are considered as maps $\mathcal{A} \to E$ rather than as maps $\mathcal{A} \to T\mathcal{A}$, and one defines $[X, Y](J) = DY(J)X(J) - DX(J)Y(J)$. One has to show that the vector obtained by this definition is indeed tangent to $\mathcal{A}$ in J. To see this, we shall define projections $\Pi(J) : E \to T_J\mathcal{A}$. Then, differentiating the relation $\Pi(J)X(J) \equiv X(J)$, which holds for every tangent vector field X, with respect to J at J in direction $Y(J)$, one obtains:

$$(D\Pi(J)Y(J))\, X(J) + (\Pi(J) - id)\, DX(J)\, Y(J) \equiv 0 \ .$$

Similarly, with the roles of X and Y interchanged:

$$(D\Pi(J)X(J))\, Y(J) + (\Pi(J) - id)\, DY(J)\, X(J) \equiv 0 \ .$$

We claim that the respective first terms of these equations are equal. Hence

$$(\Pi(J) - id)[X, Y](J) \equiv 0$$

by subtracting the equations, so $[X, Y]$ is a tangent vector field.

The task of constructing $\Pi(J)$ is effected in coordinates, and there are quite explicit coordinates available in the present situation: X and Y being vector fields on $\mathcal{A}$, for each $J \in \mathcal{A}$, $X(J)$ and $Y(J)$ are trace free 1-1 tensors on M which are symmetric with respect to $g(J)$ according to lemma 1.3.5. Then $DY(X)(J)$ will still be a trace free 1-1 tensor on M, but need no longer be symmetric with respect to $g(J)$. For given J, define the projection map $\Pi(J)$ by

$$\Pi(J) : Z \mapsto \frac{1}{2}(Z + Z^*) \ , \tag{4.1}$$

defined for all 1-1 tensors Z on M, where Z^* denotes the adjoint of Z with respect to $g(J) = \tilde\Psi(J)$, the Poincaré metric associated to J (cf. theorems 1.3.4 and 1.6.2).

In coordinates, the tensor Z is represented by a matrix (precisely: a matrix valued function on M), again called Z. Similarly let G be the matrix of the metric. Then denoting by Z^t the transpose, one easily gets the following forms for Π and its derivative in direction Y:

$$\begin{aligned}
\Pi(J) \ : \ Z \ &\mapsto \ \tfrac{1}{2}(Z + G^{-1}Z^t G) \\
D\Pi(J)Y \ : \ Z \ &\mapsto \ \tfrac{1}{2}\left(-G^{-1}(DG(J)Y)G^{-1}Z^t G + G^{-1}Z^t(DG(J)Y)\right) \ .
\end{aligned} \tag{4.2}$$

Using equation (2.6) on page 57 for $DG(J)Y$ we see that the term ρg appearing there cancels in the above formula; moreover using conformal coordinates $G = p \cdot I$ on M, that $Z^t = Z$, and $h = (YJ)_\flat$ is represented by the matrix $p \cdot YJ$. Thus

$$D\Pi(J)Y : Z \mapsto \frac{1}{2}(-p^{-1}pYJZ + p^{-1}ZpYJ) = \frac{1}{2}J(YZ + ZY) \ . \tag{4.3}$$

All this holds for a general 1-1 tensor Z on M, especially for $Z = X(J)$, where X is a vector field on $\mathcal{A}$. The symmetry of the last expression verifies the above claim.

The role the Nijenhuis tensor plays is contained in

Theorem 4.1.1 *If* $N(J) \equiv 0$ *for an almost complex structure* J *on a finite dimensional manifold* N, *then* J *is integrable, i.e. there exists a complex structure whose associated almost complex structure is* J.

This famous theorem, due to Newlander and Nirenberg, is proved in [80]. We are not going to repeat the proof here. At the end of this chapter it will even turn out that we can avoid using the theorem for our purposes by giving complex coordinates explicitly.

We cannot apply this theorem to $\mathcal{A}$ because $\mathcal{A}$ is not finite dimensional. Even though an analogous theorem holds for Hilbert manifolds we shall not try to apply such a generalized theorem to, say, $\mathcal{A}^s$. We will rather show that Φ on $\mathcal{A}$ with $N(\Phi) \equiv 0$ passes to an almost complex structure $\hat{\Phi}$ on the quotient $T(M) = \mathcal{A}/\mathcal{D}_0$ and satisfies $N(\hat{\Phi}) \equiv 0$ there. This motivates the following

Definition 4.1.4 Let $(\pi, \mathcal{G}, P, \Sigma)$ be a (weak) principal $\mathcal{G}$-bundle in the sense of section 2.5. It is said to be an almost complex $\mathcal{G}$-bundle, if

(i) P has an almost complex structure J

(ii) J is $\mathcal{G}$-invariant, i.e. $DR_a^{-1} J_{p \cdot a} DR_a = J_p$ where $R_a : p \mapsto p \cdot a$ is the right translation by a

(iii) J takes vertical vectors to vertical vectors, i.e.

$$V_p \in \ker D\pi_p \Rightarrow J_p V_p \in \ker D\pi_p \ .$$

Theorem 4.1.2 *Under the conditions of definition 4.1.4,* J *induces an almost complex structure* J_Σ *on* Σ. *If* $N(J) \equiv 0$, *then* $N(J_\Sigma) \equiv 0$.

PROOF: Let $X_q \in T_q\Sigma$ and choose $p \in \pi^{-1}(q)$ and $\tilde{X}_p$ such that $D\pi_p \tilde{X}_p = X_q$. Then define $(J_\Sigma X)_q := D\pi_p J_p \tilde{X}_p$. Thus, we have to check that the definition is independent of the choices $\tilde{X}_p$ and p. The first follows from condition (iii): Let $\tilde{X}'_p$ be another vector such that $D\pi_p \tilde{X}'_p = X_q$. Then $\tilde{X}'_p - \tilde{X}_p$ is vertical, therefore so is $J_p(\tilde{X}'_p - \tilde{X}_p)$ by (iii). This means $D\pi_p J_p \tilde{X}'_p - D\pi_p J_p \tilde{X}_p = 0$.

By a similar calculation the definition does not depend on the choice of p due to condition (ii).

Now, if Z_1, Z_2 are two $\mathcal{G}$-invariant vector fields on P, their pushed forward vector fields $\pi_* Z_i$ can be defined by $(\pi_* Z)_{\pi(p)} = D\pi_p Z_p$, and we have the identity

$$\pi_* \mathbf{N}(J)(Z_1, Z_2) = \mathbf{N}(J_\Sigma)(\pi_* Z_1, \pi_* Z_2) \ .$$

Hence $\mathbf{N}(J_\Sigma)$ is zero if $\mathbf{N}(J)$ is. ∎

We have already seen (Theorem 1.1.1) that in two dimensions every almost complex structure gives rise to a complex structure. Hence, the Nijenhuis tensor has to vanish identically in two dimensions. We can also check this explicitly:

Since (X, JX) is a basis, let $Y = aX + bJX$. Then $\mathbf{N}(X, Y) = \mathbf{N}(X, JX)$ by skew symmetry.

$$\begin{aligned}
\tfrac{1}{2}\mathbf{N}(J)(X, JX) &= [JX, J^2 X] - [X, JX] - J[JX, JX] - J[X, J^2 X] \\
&= -[JX, X] - [X, JX] - 0 + 0 \\
&= 0 \ .
\end{aligned}$$

Let us next show

Theorem 4.1.3 *The natural almost complex structure Φ on $\mathcal{A}$ makes $(\pi, \mathcal{D}_0, \mathcal{A}, T(M))$ an almost complex principal $\mathcal{D}_0$-bundle, and $\mathbf{N}(\Phi) \equiv 0$.*

Remark 4.1.2 *If M is any even dimensional orientable manifold, then the manifold $\mathcal{A}$ of all almost complex structures on M also has an almost complex structure. For this, we did not use the fact that M is two dimensional.*

PROOF:

(1) "Φ maps vertical to vertical": Consider a point $J \in \mathcal{A}$. It is by definition a tensor field on M. Since M is two dimensional, $\mathbf{N}(J)$ vanishes identically on M. Let X, Y

be vector fields on M. Then

$$
\begin{aligned}
0 &= \tfrac{1}{2}\mathbf{N}(J)(X,Y) = L_{JX}(JY) - L_X Y - JL_{JX}Y - JL_X(JY) \\
&= (L_{JX}J)Y + JL_{JX}Y - L_X Y - JL_{JX}Y - J(L_X J)Y - J^2 L_X Y \\
&= (L_{JX}J - JL_X J)Y \ .
\end{aligned}
$$

Since this holds for all Y, we have

$$
L_{JX}J = JL_X J \qquad \text{for all } X \ . \tag{4.4}
$$

Remark 4.1.3 *This equation gives a new interpretation of the vanishing of the Nijenhuis tensor, namely that every vertical vector $L_X J \in T_J \mathcal{A}$ is mapped under the almost complex structure Φ on $\mathcal{A}$ into a vertical vector (in fact $L_{JX}J$) again. Thus $\mathbf{N}$ can be thought of as the obstruction of preserving the vertical for the natural almost complex structure Φ.*

(2) "Φ is $\mathcal{D}$-invariant": Let $O_f(J) := f^*J$ and note that $DO_f(J)H = f^*H$. For the $\mathcal{D}$-invariance of Φ we have to show that

$$
DO_f(J)^{-1}\Phi_{O_f(J)}DO_f(J)H = \Phi_J(H) \ .
$$

This is equivalent to

$$
(f^*)^{-1}(f^*Jf^*H) = JH \ ,
$$

which is true because $f^*Jf^*H = f^*(JH)$.

(3) "$\mathbf{N}(\Phi) \equiv 0$":

Let Z, W be vector fields on $\mathcal{A}$ and for the moment denote by $DW(Z)$ the function $J \mapsto DW(J)Z(J)$. Then we have $DJ(Y) = Y$, the derivative of the identity. Together with the product rule, this yields the following calculation:

$$
\begin{aligned}
\tfrac{1}{2}\mathbf{N}(\Phi)(Z,W) &= [\Phi Z, \Phi W] - [Z, W] - \Phi[\Phi Z, W] - \Phi[Z, \Phi W] \\
&= D(JW)(JZ) - D(JZ)(JW) - DW(Z) + DZ(W) \\
&\quad - J\,DW(JZ) + J\,D(JZ)(W) - J\,D(JW)(Z) + J\,DZ(JW) \\
&= (JZ)W + J\,DW(JZ) - (JW)Z - J\,DZ(JW) \\
&\quad - DW(Z) + DZ(W) \\
&\quad - J\,DW(JZ) + JWZ + J^2\,DZ(W) \\
&\quad - JZW - J^2\,DW(Z) + J\,DZ(JW) \\
&= 0 \ . \qquad\qquad\qquad\qquad\qquad\qquad\qquad\qquad \blacksquare
\end{aligned}
$$

If we had an analog of theorem 4.1.1 in infinite dimensions we could now say that the almost complex structure on $\mathcal{A}$ is integrable. But theorem 4.1.1 does guarantee that the almost complex structure $\mathring{\Phi}$ on $\mathcal{A}/\mathcal{D}_0$ which is induced by Φ according to theorem 4.1.2, is integrable. We therefore have a complex structure on Teichmüller space.

Remark 4.1.4 *Note that if we choose J to correspond to multiplication by $+i$ in a coordinate chart on M as we did in the beginning of section 1.1 (i.e. $J = \begin{bmatrix} 0 & -1 \\ 1 & 0 \end{bmatrix}$ rather than $\begin{bmatrix} 0 & 1 \\ -1 & 0 \end{bmatrix}$ in conformal coordinates), then the corresponding almost complex structure on Teichmüller space means multiplication by $-i$ on the level of holomorphic quadratic differentials: in fact let $g_{kl} = \lambda \delta_{kl}$ and consider the tangent vector*

$$h^{TT} = u\, dx^2 - u\, dy^2 - 2v\, dx\, dy = \operatorname{Re}\left((u+iv)dz^2\right)$$

to Teichmüller space, so raising an index, $H = \lambda \begin{bmatrix} u & -v \\ -v & -u \end{bmatrix}$.

Then $JH = \begin{bmatrix} v & u \\ u & -v \end{bmatrix}$, which corresponds to the tangent vector $\operatorname{Re}\left((v-iu)dz^2\right)$. If we chose J instead to be multiplication by $-i$ on M, then the induced almost complex structure on $T(M)$ would correspond to multiplication by i on the holomorphic quadratic differentials.

One can easily see that the following definition of holomorphy agrees with the usual one:

Definition 4.1.5 Let (M, J) and $(N, \hat{J})$ be complex manifolds with almost complex structures J and $\hat{J}$ respectively. Then a map $f : M \to N$ is called holomorphic, if $Df\, J = \hat{J}\, Df$. By this we mean: $Df(x) \circ J(x) = \hat{J}(f(x)) \circ Df(x)$ for all $x \in M$.

We have already seen that $\mathcal{D}$ acts on $\mathcal{A}$, and that Φ is $\mathcal{D}$-invariant, i.e.

$$\Phi DO_f = DO_f \Phi\ .$$

We can paraphrase this equation by saying that the action O_f of $\mathcal{D}$ on $\mathcal{A}$ is formally holomorphic. We use the term formally holomorphic because one condition of definition 4.1.5 is not satisfied; $\mathcal{A}$ is not a complex manifold, but an almost complex infinite dimensional manifold whose almost complex structure is formally integrable (i.e. $N(\Phi) = 0$). Similarly, $\pi : \mathcal{A} \to T(M)$ is "holomorphic" since by definition $\hat{\Phi} D\pi = D\pi \Phi$. We can "factor out" $\mathcal{D}_0$ from this remark to get something finite dimensional which is no longer only formal:

Theorem 4.1.4 $\mathcal{D}/\mathcal{D}_0$ acts on $T(M) = \mathcal{A}/\mathcal{D}_0$ as a group of holomorphic mappings.

PROOF: For $f \in \mathcal{D}$, $[J] \in \mathcal{A}/\mathcal{D}_0$, consider the map $[J] \mapsto [f^*J]$. This is well-defined, i.e. $[f^*J]$ depends only on $[J]$, but not on J, because for $J_1 = h^*J_2$ with $h \in \mathcal{D}_0$ we have:

$$f^*J_1 = f^*h^*J_2 = \left(f^*h^*(f^{-1})^*\right) f^*J_2$$

with $f^*h^*(f^{-1})^* \in \mathcal{D}_0$. Thus $\mathcal{D}$ acts on $\mathcal{A}/\mathcal{D}_0$; but obviously the normal subgroup $\mathcal{D}_0$ acts trivially on $\mathcal{A}/\mathcal{D}_0$, hence the action passes to an action of the quotient $\mathcal{D}/\mathcal{D}_0$. We have to show that $\hat{O}_f : [J] \mapsto [f^*J]$ is a holomorphic map, i.e. that

$$\hat{\Phi} D\hat{O}_f = D\hat{O}_f \hat{\Phi} \ .$$

This follows from the following three equations:

(1) $D\hat{O}_f D\pi = D\pi DO_f$
    (the differential version of $\hat{O}_f \circ \pi = \pi \circ O_f$)

(2) $D\pi\Phi = \hat{\Phi}D\pi$
    (the definition of $\hat{\Phi}$; this also says that π is holomorphic)

(3) $DO_f\Phi = \Phi \circ DO_f$
    (the $\mathcal{D}$-equivariance of Φ or formal holomorphy of O_f) .

Now:

$$D\hat{O}_f \hat{\Phi} D\pi \stackrel{(2)}{=} D\hat{O}_f D\pi \Phi \stackrel{(1)}{=} D\pi DO_f \Phi \stackrel{(3)}{=} D\pi \Phi DO_f \stackrel{(2)}{=} \hat{\Phi} D\pi DO_f \stackrel{(3)}{=} \hat{\Phi} D\hat{O}_f D\pi \ .$$

The claim follows since $D\pi$ is surjective. ∎

Remark 4.1.5 *It is a deep and beautiful theorem of H. Royden [94] that for genus$(M) \geq 3$, $\mathcal{D}/\mathcal{D}_0$ is the full group of complex automorphisms of $T(M)$. We do not give a proof of this important result in these notes.*

Clearly, since π is holomorphic, local holomorphic sections exist. We have already seen above that global sections exist. Cliff Earle [19] has shown, however, that no global holomorphic section exists.

4.2   Abresch-Fischer Holomorphic Coordinates for $\mathcal{A}$

We already know that "formally" $\mathcal{A}$ is a complex manifold in that the Nijenhuis tensor vanishes. In this section we introduce the explicit holomorphic coordinates on $\mathcal{A}$ discovered by Uwe Abresch and Arthur Fischer. They are, in fact, holomorphic coordinates for the space of almost complex structures over any almost complex manifold. Remarkably, in two dimensions they turn out to be global coordinates. Later, we shall see how they induce complex coordinates for $T(M)$.

Let U be an appropriate neighbourhood of 0 in $T_{J_0}\mathcal{A}$. Define

$$\psi_{J_0} : U \to \mathcal{A}$$

by

$$\psi_{J_0}(H) := (I + H)J_0(I + H)^{-1} =: J \ . \tag{4.5}$$

It is clear that $J^2 = -I$ iff $J_0^2 = -I$, which says that the range of ψ is in $\mathcal{A}$. A straightforward algebraic exercise shows that the inverse $\psi^{-1} : \psi(U) \to T_J \mathcal{A}$ of ψ is given by

$$\psi_{J_0}^{-1}(J) = (J - J_0)(J + J_0)^{-1} \ . \tag{4.6}$$

When not necessary, we choose to omit the subscript J_0 from ψ_{J_0}.

We claim that

$$\psi^*(\Phi) = \Phi_{J_0} \ , \tag{4.7}$$

where Φ is the almost complex structure on $\mathcal{A}$ ($\Phi_J(K) = JK$) and $\Phi_{J_0} : T_{J_0}\mathcal{A} \to T_{J_0}\mathcal{A}$ is the fixed linear almost complex structure on $T_{J_0}\mathcal{A}$.

Relation (4.7) says that ψ is a holomorphic coordinate chart for Φ. To verify (4.7) let $\dot{J} \in T_{J_0}\mathcal{A}$ be arbitrary. We must show that

$$D\psi(H)^{-1}\Phi_J D\psi(H)\dot{J} = \Phi_{J_0}(\dot{J}) = J_0\dot{J} \tag{4.8}$$

which is equivalent to (4.7).

On the open set of invertible 1-1 tensors the derivative of the map $G \mapsto G^{-1}$ is given by $K \mapsto -G^{-1}KG^{-1}$. Using this fact we see that

$$
\begin{aligned}
D\psi(H)\dot{J} &= \dot{J}J_0(I + H)^{-1} - (I + H)J_0(I + H)^{-1}\dot{J}(I + H)^{-1} \\
&= \dot{J}J_0(I + H)^{-1} - J\dot{J}(I + H)^{-1} \ .
\end{aligned} \tag{4.9}
$$

Therefore

$$
\begin{aligned}
JD\psi(H)\dot{J} &= J\dot{J}J_0(I + H)^{-1} + \dot{J}(I + H)^{-1} \\
&= -JJ_0\dot{J}(I + H)^{-1} + \dot{J}(I + H)^{-1} \\
&= (I - JJ_0)\dot{J}(I + H)^{-1} \\
&= -J(J + J_0)\dot{J}(I + H)^{-1} \ .
\end{aligned} \tag{4.10}
$$

From (4.6) we see that the derivative $D\psi(H)^{-1}$ is given by

$$
\begin{aligned}
D\psi^{-1}(J)K &= K(J + J_0)^{-1} - (J - J_0)(J + J_0)^{-1}K(J + J_0)^{-1} \\
&= K(J + J_0)^{-1} - HK(J + J_0)^{-1} \\
&= (I - H)K(J + J_0)^{-1} \ .
\end{aligned} \tag{4.11}
$$

Consequently

$$
\begin{aligned}
D\psi(H)^{-1}\Phi_J D\psi(H)\dot{J} &= \\
&= D\psi(H)^{-1}J D\psi(H)\dot{J} \\
&= (I - H)\left(-J(J + J_0)\dot{J}(I + H)^{-1}\right)(J + J_0)^{-1} \\
&= -(I - H)\left(J(J + J_0)\dot{J}\right)\left((J + J_0)(I + H)\right)^{-1} \\
&= -(I - H)\left((I + H)J_0(I + H)^{-1}(J + J_0)\dot{J}\right)\left((J + J_0)(I + H)\right)^{-1} \ .
\end{aligned}
\tag{4.12}
$$

However

$$
\begin{aligned}
(I - H)(I + H)J_0(I + H)^{-1} &= (I + H)(I - H)J_0(I + H)^{-1} \\
&= (I + H)J_0 = J_0(I - H) \ ,
\end{aligned}
$$

and so the last line of (4.12) is equal to

$$
- J_0(I - H)(J + J_0)\dot{J}\left((J + J_0)(I + H)\right)^{-1} \ .
\tag{4.13}
$$

But

$$
\begin{aligned}
(I - H)(J + J_0) &= \left(I - (J - J_0)(J + J_0)^{-1}\right)(J + J_0) \\
&= \left((J + J_0) - (J - J_0)\right) \\
&= 2J_0 \ .
\end{aligned}
$$

Moreover the term

$$
(J + J_0)(I + H) = (J + J_0)\left(I + (J - J_0)(J + J_0)^{-1}\right) \ ,
\tag{4.14}
$$

and since

$$
(J + J_0)(J - J_0) = -(J - J_0)(J + J_0)
$$

(4.14) is equal to

$$
\left((J + J_0) - (J - J_0)\right) = 2J_0 \ .
$$

Noting that $J_0^{-1} = -J_0$ we see that

$$
\left((J + J_0)(I + H)\right)^{-1} = -\frac{1}{2}J_0 \ .
$$

Returning to the last line (4.13) of our computation for $D\psi(H)^{-1}J D\psi(H)$ we see that

$$
\begin{aligned}
D\psi(H)^{-1}J D\psi(H)\dot{J} &= -J_0(2J_0)\dot{J}\left(-\tfrac{1}{2}J_0\right) \\
&= -\dot{J}J_0 \\
&= J_0\dot{J} \ .
\end{aligned}
$$

Therefore ψ is a complex coordinate chart for the almost complex structure Φ. Clearly such a chart can be found about each $J_0 \in \mathcal{A}$ and we have shown:

Theorem 4.2.1 *Abresch-Fischer coordinates ψ (see formula (4.5)) are holomorphic complex coordinates for $\mathcal{A}$. Therefore $\mathcal{A}$ is a complex manifold.*

Remark 4.2.1 *In two dimensions, it is elementary linear algebra to see that ψ is surjective.*

In introducing Abresch-Fischer coordinates above, we allowed U to be an appropriate neighbourhood of 0 in $T_{J_0}\mathcal{A}$ so that $H \in U$ iff $I + H$ is invertible. We now show that U can be characterized as the unit ball in $T_{J_0}\mathcal{A}$ in the L^∞ norm, i.e.

Theorem 4.2.2 *The space $\mathcal{A}$ of all almost complex structures on a closed Riemannian surface M is holomorphically equivalent to the open unit ball $B^\infty(J_0) \subset T_{J_0}\mathcal{A}$ in the L^∞ norm, i.e.*

$$B^\infty(J_0) = \left\{ H \in T_{J_0}\mathcal{A} \,\Big|\, \sup_x |H_x| < 1 \right\}$$

where

$$|H_x| := \sup_{v_x \in T_x M} \left\{ |g_0(x)(H_x v_x, v_x)| \,\Big|\, g_0(x)(v_x, v_x) \le 1 \right\}$$

and $g_0 = g_0(J_0) := \check{\Psi}(J_0)$, the Poincaré metric associated to J_0.

PROOF: Recall that each H_x is trace free and symmetric with respect to $g_0(x)$ (lemma 1.3.5). Clearly, ψ is defined on $B^\infty(J_0)$ since every $H \in B^\infty(J_0)$ has the property that $I + H$ is invertible. Now suppose $\psi(H) = J \in \mathcal{A}$. We wish to show that $H \in B^\infty(J_0)$. From formula (4.6) we have

$$H = (J - J_0)(J + J_0)^{-1} \tag{4.15}$$

or

$$H = J_0(J_0^{-1}J - I)(I + J_0^{-1}J)^{-1}J_0^{-1} \ ,$$

where $J_0^{-1} = -J_0$. Clearly, $|H_x| = |B_x|$,

$$B_x := (J_0(x)^{-1}J(x) - I)(I + J_0(x)^{-1}J(x))^{-1} \ . \tag{4.16}$$

Let us work in conformal coordinates where $(g_0)_{ij} = \lambda\delta_{ij}$. Then $J_0 = \left(\begin{smallmatrix} 0 & -1 \\ 1 & 0 \end{smallmatrix}\right)$. Let $J = \left(\begin{smallmatrix} a & b \\ c & -a \end{smallmatrix}\right)$. Since $J^2 = -I$,

$$a^2 + bc = -1 \ . \tag{4.17}$$

Simple algebraic computations show that

$$(I + J_0^{-1}J)^{-1} = \frac{1}{2 + c - b}\left(\begin{array}{cc} 1 - b & a \\ a & 1 + c \end{array} \right) \tag{4.18}$$

and

$$J_0^{-1}J - I = \begin{pmatrix} c-1 & -a \\ -a & -b-1 \end{pmatrix} . \tag{4.19}$$

Moreover,

$$B = \frac{1}{2+c-b}\begin{pmatrix} c+b & -2a \\ -2a & -c-b \end{pmatrix} ,$$

and the eigenvalues of B are

$$\pm \frac{\sqrt{(c-b)^2 - 4}}{2+c-b} = \pm\sqrt{\frac{c-b-2}{c-b+2}} . \tag{4.20}$$

Since J is in the same component of the set of all 2×2 matrices whose square is $-I$ as is J_0, and since by (4.17), $bc < 0$ for all these matrices, it follows that $b < 0$ and $c > 0$. Thus for any x,

$$|B_x| \le \sqrt{|(c-b-2)/(c-b+2)|} < 1 ,$$

finishing the proof. ∎

Remark 4.2.2 *Classically, the elements of $B^\infty(J_0)$ are known as Beltrami differentials.*

4.3   Abresch-Fischer Holomorphic Coordinates for $\mathcal{T}(M)$

Let $[J_0] \in \mathcal{T}(M)$ and $H \in T_{[J_0]}\mathcal{T}(M)$ and set $\tilde{H}$ to be the horizontal $\mathcal{D}$-invariant lift of H to $T\mathcal{A}$; i.e. $\tilde{H}(J) \in T_J\mathcal{A}$ for any J with $\pi(J) = [J_0]$ and $f^*\tilde{H} = \tilde{H}$. For any $J \in \pi^{-1}[J_0]$ define

$$\hat{\psi}_{[J_0]}(H) = \pi \circ \psi_J(\tilde{H}) ,$$

$\pi : \mathcal{A} \to \mathcal{T}(M)$ the projection map, $\psi_J : T_J\mathcal{A} \to \mathcal{A}$ defined in (4.5). Since, clearly, $f^*\psi_J(\tilde{H}) = \psi_{f\cdot J}(f^*\tilde{H}) = \psi_{f\cdot J}(\tilde{H})$ and thus $\pi\psi_{f\cdot J}(\tilde{H}) = \pi\psi_J(\tilde{H})$, we see that ψ is independent of $J \in \pi^{-1}[J_0]$, and so $\hat{\psi}$ is well-defined.

Moreover, from (4.9) it follows that

$$D\psi_{J_0}(0)\dot{J} = 2JJ_0$$

and so $D\psi_{J_0}(0)$: $T_{J_0}\mathcal{A} \to T_{J_0}\mathcal{A}$ is an isomorphism, which since it is multiplication on the right by $2J_0$ preserves both the horizontal and vertical vectors in $T_{J_0}\mathcal{A}$. This immediately implies that $D\hat{\psi}_{[J_0]}(0) : T_{[J_0]}\mathcal{T}(M) \to T_{[J_0]}\mathcal{T}(M)$ is an isomorphism and therefore by the implicit function theorem $\hat{\psi}_{[J_0]}$ restricted to some small neighbourhood of the origin in $T_{[J_0]}\mathcal{T}(M)$ is a diffeomorphism onto its image. It remains to show that $\hat{\psi}_{[J_0]}$ is holomorphic.

But $\hat{\psi} = \pi \circ \psi$ and both π and ψ are holomorphic. Constructing $\hat{\psi}_{[J_0]}$ for each $[J_0] \in \mathcal{T}(M)$ provides a holomorphic coordinate atlas for $\mathcal{T}(M)$.

Remark 4.3.1 *One might ask whether Abresch-Fischer coordinates could also be global coordinates for* $\mathcal{T}(M)$ *as they are for* $\mathcal{A}$*. The answer is no; if they were global, one could then produce a global holomorphic section to the principal bundle* $\pi : \mathcal{A} \rightarrow \mathcal{T}(M)$*, which, as we remarked at the end of section 4.1, is impossible by a result of Cliff Earle.*

5 Properties of the Weil-Petersson Metric

5.1 The Weil-Petersson Metric is Kähler

We open this section with some abstract theorems on principal fibre bundles which we later apply to Teichmüller space $\mathcal{T}(M)$. We begin by considering the notion of a weak Riemannian principal bundle $(\pi, \mathcal{G}, P, \Sigma)$, $\mathcal{G}$ a Lie group, P a (Fréchet) manifold with $\Sigma = P/\mathcal{G}$.

Definition 5.1.1 A *Riemannian principal fibre bundle* $(\pi, \mathcal{G}, P, \Sigma)$ is a principal $\mathcal{G}$-bundle with a $\mathcal{G}$-invariant Riemannian structure G on the total space P. Thus for each $p \in P$, $G(p) : T_p P \times T_p P \to \mathbb{R}$ and $p \mapsto G(p)$ is smooth. For $p \in P$ there is a vertical subspace $\mathcal{V}(p) \subset T_p P$ given by

$$\mathcal{V}(p) = \mathrm{Ker}\, D\pi(p)$$

the kernel of the derivative of the projection. We shall further assume that $\mathcal{V}(p)$ has an orthogonal complement with respect to G, which we call the *horizontal* subspace $\mathcal{H}(p)$ at p. Note however that the topology induced on P by G *can be weaker* than the given topology.

Thus

$$T_p P = \mathcal{H}(p) \oplus \mathcal{V}(p) \tag{5.1}$$

and the decomposition (5.1) is $\mathcal{G}$-orthogonal. Then exactly as with the L^2-metric on $(\pi, \mathcal{D}_0, \mathcal{M}_{-1}, \mathcal{M}_{-1}/\mathcal{D}_0)$ we have

Theorem 5.1.1 *Let $(\pi, \mathcal{G}, P, \Sigma)$ be a C^∞ Riemannian principal fibre bundle with Riemannian structure G. Then G naturally induces a Riemannian structure G_Σ on Σ.*

The induced structure G_Σ on Σ is defined as for the case of $(\pi, \mathcal{D}_0, \mathcal{M}_{-1}, \mathcal{M}_1/\mathcal{D}_0)$. For $x \in \Sigma$, $X_x, Y_x \in T_x\Sigma$ we set

$$G_\Sigma(x)(X_x, Y_x) = G(p)(\tilde{X}_p, \tilde{Y}_p) \ ,$$

where p is any point with $\pi(p) = x$, and $\tilde{X}_p, \tilde{Y}_p \in H_p$ are the unique horizontal lifts of X_x, Y_x respectively. Thus, $D\pi(p)\tilde{X}_p = X_x$, and similarly for $\tilde{Y}_p$ and Y_x.

One again readily checks that the equivariance of G under the group $\mathcal{G}$ implies that G_Σ is well-defined and is a Riemannian metric for Σ.

Theorem 5.1.2 $(G_\Sigma(X, Y)) \circ \pi = G(\tilde{X}, \tilde{Y})$.

If X and Y are C^∞ vector fields on Σ with unique horizontal lifts $\tilde{X}, \tilde{Y}$, then $\tilde{X}$ and $\tilde{Y}$ are $\mathcal{G}$-invariant vector fields on P.

Let $(\pi, \mathcal{G}, \Phi, P, \Sigma)$ an almost complex principal bundle admitting an almost complex structure Φ. Then we know that Φ induces a natural almost complex structure $\hat{\Phi}$ on Σ

$$\hat{\Phi}(x)X_x = D\pi(p)\Phi(p)\tilde{X}_p \ ,$$

where p is any point with $\pi(p) = x$, and $\tilde{X}_p$ is *any* vector in $T_p p$ with $D\pi(p)\tilde{X}_p = X_x$. Then we have

Theorem 5.1.3 *Let $(\pi, \mathcal{G}, \Phi, P, \Sigma)$ be a Riemannian almost complex principal fibre bundle with Riemannian structure G and almost complex structure Φ. If G is Hermitian with respect to Φ, then G_Σ is a Hermitian metric with respect to $\hat{\Phi}$.*

PROOF:

$$G_\Sigma(x) \left(\hat{\Phi}(x)X_x, \hat{\Phi}(x)Y_x \right) = G_\Sigma(x) \left(D\pi(p)\Phi(p)\tilde{X}_p, D\pi(p)\Phi(p)\tilde{Y}_p \right) \ ,$$

where $\tilde{X}_p, \tilde{Y}_p$ are the unique horizontal lifts of X_x, Y_x to $p \in \pi^{-1}(x)$. Since G is Hermitian with respect to Φ, it follows that Φ preserves the horizontal as well as vertical subspaces. Then by definition, the above quantity is equal to

$$G(p) \left(\Phi(p)\tilde{X}_p, \Phi(p)\tilde{Y}_p \right) \ .$$

Since G is Hermitian with respect to J, by definition this is equal to

$$G(p)(\tilde{X}_p, \tilde{Y}_p) = G_\Sigma(x)(X_x, Y_x) \ ,$$

which concludes theorem 5.1.3.   ∎

Definition 5.1.2 Let Σ be a complex Riemannian manifold with almost complex structure $\hat{\Phi}$ and Hermitian metric G_Σ. We say that G_Σ is Kähler if the Kähler two-form Ω_Σ, defined by

$$\Omega_\Sigma : T\Sigma \times T\Sigma \to \mathbb{R}; \quad \Omega_\Sigma(X_x, Y_x) = G_\Sigma(\hat{\Phi}X_x, Y_x) \ ,$$

is closed, i.e. $d\Omega_\Sigma = 0$.

We now have the following result:

Theorem 5.1.4 *Let $(\pi, \mathcal{G}, \Phi, P, \Sigma)$ be an almost complex Riemannian principal $\mathcal{G}$-bundle with almost complex structure Φ and Riemannian structure G. Let G_Σ and $\hat{\Phi}$ denote the induced structures on Σ with Ω and Ω_Σ the corresponding Kähler forms. For $X, Y, Z \in \mathcal{X}(\Sigma)$ (the vector fields on Σ) we have*

$$d\Omega_\Sigma(X, Y, Z) \circ \pi = d\Omega(\tilde{X}, \tilde{Y}, \tilde{Z}) \ ,$$

where $\tilde{X}, \tilde{Y}, \tilde{Z}$ are the unique horizontal lifts of X, Y, Z to vector fields on P.

PROOF: For this we use the well known formula for $d\Omega$ and $d\Omega_\Sigma$ (see [64]):

$$d\Omega(\tilde{X}, \tilde{Y}, \tilde{Z}) \; = \; \frac{1}{3}\Big\{\tilde{X}\left(\Omega(\tilde{Y}, Z)\right) + \tilde{Y}\left(\Omega(\tilde{Z}, \tilde{X})\right) + \tilde{Z}\left(\Omega(\tilde{X}, \tilde{Y})\right) \qquad (5.2)$$
$$- \Omega\left([\tilde{X}, \tilde{Y}], \tilde{Z}\right) - \Omega\left([\tilde{Y}, \tilde{Z}], \tilde{X}\right) - \Omega\left([\tilde{Z}, \tilde{X}], \tilde{Y}\right)\Big\} \ ,$$

where $[\cdot, \cdot]$ denotes the Lie bracket and, for φ a smooth real valued function on P, and $X(\varphi) = D\varphi(X)$.

From the construction of G_Σ and $\hat{\Phi}$ it follows that

$$(\Omega_\Sigma(Y, Z)) \circ \pi = \Omega(\tilde{Y}, \tilde{Z}) \ .$$

Thus

$$\tilde{X}\left(\Omega(\tilde{Y}, \tilde{Z})\right) \; = \; \tilde{X}\left(\Omega_\Sigma(Y, Z) \circ \pi\right)$$
$$= \; D\left(\Omega_\Sigma(Y, Z)\right) \circ D\pi(\tilde{X})$$
$$= \; \left(D\left((\Omega_\Sigma(Y, Z)) \circ X\right) \circ \pi\right.$$
$$= \; \left(X\left(\Omega_\Sigma(Y, Z)\right)\right) \circ \pi \ ,$$

and similarly, for the second two terms in (5.2). For a $\mathcal{G}$-invariant vector field $\tilde{X}$ on P, denote by $\pi_*(\tilde{X})$ the pushed down vector field on Σ,

$$\pi_*(\tilde{X})(x) = D\pi(\tilde{X}) \circ \pi^{-1}(x) \ .$$

If $\tilde{X}$ is the lift of X, then $\pi_*\tilde{X} = X$. Since

$$\pi_*[\tilde{X},\tilde{Y}] = [\pi_*\tilde{X},\pi_*\tilde{Y}] = [X,Y] \ ,$$

if follows that the horizontal component of $[\tilde{X},\tilde{Y}]$, namely $[\tilde{X},\tilde{Y}]^H$, is a horizontal lift of $[X,Y]$, i.e.

$$[\widetilde{X,Y}] = [\tilde{X},\tilde{Y}]^H \ .$$

Thus, if $[\tilde{X},\tilde{Y}]^V$ denotes the vertical part of $[\tilde{X},\tilde{Y}]$,

$$[\tilde{X},\tilde{Y}] = [\widetilde{X,Y}] + [\tilde{X},\tilde{Y}]^V \ .$$

Consequently,

$$\Omega\left([\tilde{X},\tilde{Y}],\tilde{Z}\right) = \Omega\left([\widetilde{X,Y}],\tilde{Z}\right) + \Omega\left([\tilde{X},\tilde{Y}]^V,\tilde{Z}\right) \ .$$

It is easy to see that, since $\tilde{Z}$ is horizontal, $\Omega\left([\tilde{X},\tilde{Y}]^V,\tilde{Z}\right) = 0$. Thus

$$\Omega\left([\tilde{X},\tilde{Y}],\tilde{Z}\right) = \Omega_\Sigma\left([X,Y],Z\right) \circ \pi \ ,$$

and similarly, for the last three terms in (5.2). Using exactly the same formula for $d\Omega_\Sigma$ as for $d\Omega$ and substituting these results in (5.2) we obtain the conclusion of Theorem 5.1.4. ∎

We immediately have

Corollary 5.1.5 *Let* $(\pi,\mathcal{G},\Phi,P,\Sigma)$ *be an almost complex Riemannian principal* $\mathcal{G}$-*bundle and* Ω,Ω_Σ *the Kähler forms on* P *and* Σ *induced by the almost complex and Riemannian structures on* P. *Then* Σ *is Kähler if* $d\Omega(\tilde{X},\tilde{Y},\tilde{Z}) = 0$ *for horizontal vectors* $\tilde{X},\tilde{Y},\tilde{Z}$.

PROOF: By (5.1.4)

$$d\Omega_\Sigma(X,Y,Z) \circ \pi = d\Omega(\tilde{X},\tilde{Y},\tilde{Z}) \ .$$

∎

Let us now return to our study of the almost complex Riemannian principal bundle $(\pi,\mathcal{D}_0,\mathcal{A},\mathcal{A}/\mathcal{D}_0)$. Again let Φ be the almost complex structure on $\mathcal{A}$, which at $J \in \mathcal{A}$ is multiplication by J on $T_J\mathcal{A}$, let $\langle\langle\cdot,\cdot\rangle\rangle$ be the L^2-metric on $\mathcal{A}$ and $\langle\cdot,\cdot\rangle$ the induced metric on $\mathcal{A}/\mathcal{D}_0 = T(M)$.

Denote by Ω and Ω_T the corresponding Kähler forms on $\mathcal{A}$ and $\mathcal{T}(M)$. In order to show that $\langle \cdot, \cdot \rangle$ is Kähler we need to show that $d\Omega(\tilde{X}, \tilde{Y}, \tilde{Z}) = 0$ for all horizontal $\tilde{X}, \tilde{Y}, \tilde{Z}$.

The Kähler form on $\mathcal{A}$ is given by

$$\Omega(\tilde{Y}, \tilde{Z}) = \int_M tr(J\tilde{Y}\tilde{Z}) d\mu_{g(J)} \ . \tag{5.3}$$

Theorem 5.1.6 *For horizontal fields* $\tilde{X}, \tilde{Y}, \tilde{Z}$,

$$\tilde{X}\left(\Omega(\tilde{Y}, \tilde{Z})\right) = \int_M tr(\tilde{X}\tilde{Y}\tilde{Z}) d\mu_{g(J)} \ + \ \int_M tr\left(J\left(D\tilde{Y}(\tilde{X})\right)\tilde{Z}\right) d\mu_{g(J)}$$
$$+ \ \int_M tr\left(J\tilde{Y}\left(D\tilde{Z}(\tilde{X})\right)\right) d\mu_{g(J)} \ .$$

PROOF: Consider the map $J \mapsto g(J) = \tilde{\Psi}(J)$. Then by theorem 2.5.6 the derivative $D\tilde{\Psi}(J)$ takes horizontal vectors in $T_J\mathcal{A}$ to horizontal vectors in $T_{\tilde{\Psi}(J)}\mathcal{M}_{-1}$. Thus if $\tilde{X}(\cdot)$ is trace free and divergence free with respect to $g(J)$ then $D\tilde{\Psi}(J)\tilde{X}(J)$ is a trace free divergence free symmetric 0-2 tensor, so in particular, the trace

$$tr\left(D\tilde{\Psi}(J)\tilde{X}(J)\right) = 0 \ . \tag{5.4}$$

We already know that the derivative of the map $g \mapsto \mu_g$, μ_g the volume element of g, is given by

$$h \mapsto \left(\frac{1}{2} tr_g h\right) \mu_g \ . \tag{5.5}$$

As a consequence, applying the chain rule to (5.5) and using (5.4) we obtain

Lemma 5.1.7 *The derivative of the map* $J \mapsto \mu_{g(J)}$ *vanishes on horizontal vectors.*

In proving theorem 5.1.6 we just observe that in the expression (5.3) for Ω there are four terms which are functions of J, namely J itself, $\tilde{Y}(J)$, $\tilde{Z}(J)$ and $J \mapsto d\mu_{g(J)}$. By the previous lemma the derivative of $J \mapsto d\mu_{g(J)}$ in the direction $\tilde{X}$ is zero. The derivative of "J" in the direction $\tilde{X}$ is just $\tilde{X}$ and the derivatives of $\tilde{Y}$ and $\tilde{Z}$ in the direction $\tilde{X}$ are $D\tilde{Y}(\tilde{X})$ and $D\tilde{Z}(\tilde{X})$ respectively. Theorem 5.1.6 now follows from the product rule for differentiation.   ∎

Lemma 5.1.8 *If* $\tilde{X}, \tilde{Y}, \tilde{Z} \in T_J\mathcal{A}$ *then* $tr(\tilde{X}\tilde{Y}\tilde{Z}) = 0$.

PROOF: Since these fields anticommute with J we get

$$
\begin{aligned}
-tr(\tilde{X}\tilde{Y}\tilde{Z}) &= tr(J\tilde{X}\tilde{Y}\tilde{Z}J) \\
&= -tr(J\tilde{X}\tilde{Y}J\tilde{Z}) \\
&= tr(J\tilde{X}J\tilde{Y}\tilde{Z}) \\
&= -tr(J^2\tilde{X}\tilde{Y}\tilde{Z}) \\
&= tr(\tilde{X}\tilde{Y}\tilde{Z}) \; .
\end{aligned}
$$

∎

Lemma 5.1.9 *On horizontal fields*

$$
\tilde{X}\left(\Omega(\tilde{Y},\tilde{Z})\right) = \int_M tr\left(J\left(D\tilde{Y}(\tilde{X})\right)\tilde{Z}\right) d\mu_{g(J)} + \int_M tr\left(J\tilde{Y}\left(D\tilde{Z}(\tilde{X})\right)\right) d\mu_{g(J)} \; .
$$

PROOF: Immediate from 5.1.6 and 5.1.8. Similarly, we see that

$$
\tilde{Y}\left(\Omega(\tilde{Z},\tilde{X})\right) = \int_M tr\left(J\left(D\tilde{Z}(\tilde{Y})\right)\tilde{X}\right) d\mu_{g(J)} + \int_M tr\left(J\tilde{Z}\left(D\tilde{X}(\tilde{Y})\right)\right) d\mu_{g(J)}
$$
$$
\tilde{Z}\left(\Omega(\tilde{X},\tilde{Y})\right) = \int_M tr\left(J\left(D\tilde{X}(\tilde{Z})\right)\tilde{Y}\right) d\mu_{g(J)} + \int_M tr\left(J\tilde{X}\left(D\tilde{Y}(\tilde{Z})\right)\right) d\mu_{g(J)} \; .
$$

Moreover,

$$
\Omega\left([\tilde{X},\tilde{Y}],\tilde{Z}\right) = \int_M tr\left(J\left(D\tilde{Y}(\tilde{X}) - D\tilde{X}(\tilde{Y})\right)\tilde{Z}\right) d\mu_{g(J)} \; ,
$$
$$
\Omega\left([\tilde{Y},\tilde{Z}],\tilde{X}\right) = \int_M tr\left(J\left(D\tilde{Z}(\tilde{Y}) - D\tilde{Y}(\tilde{Z})\right)\tilde{X}\right) d\mu_{g(J)} \; ,
$$
$$
\Omega\left([\tilde{Z},\tilde{X}],\tilde{Y}\right) = \int_M tr\left(J\left(D\tilde{X}(\tilde{Z}) - D\tilde{Z}(\tilde{X})\right)\tilde{Y}\right) d\mu_{g(J)} \; .
$$

Using (5.2) and the above relationship we see that $3 \cdot d\Omega(\tilde{X},\tilde{Y},\tilde{Z})$ is given by

$$
\begin{aligned}
3 \cdot d\Omega(\tilde{X},\tilde{Y},\tilde{Z}) = \quad & \int_M tr\left(J\tilde{Y}\left(D\tilde{Z}(\tilde{X})\right)\right) d\mu_{g(J)} + \int_M tr\left(J\tilde{Z}\left(D\tilde{X}(\tilde{Y})\right)\right) d\mu_{g(J)} \\
& + \int_M tr\left(J\tilde{X}\left(D\tilde{Y}(\tilde{Z})\right)\right) d\mu_{g(J)} + \int_M tr\left(J\left(D\tilde{X}(\tilde{Y})\right)\tilde{Z}\right) d\mu_{g(J)} \\
& + \int_M tr\left(J\left(D\tilde{Y}(\tilde{Z})\right)\tilde{X}\right) d\mu_{g(J)} + \int_M tr\left(J\left(D\tilde{Z}(\tilde{X})\right)\tilde{Y}\right) d\mu_{g(J)} \; .
\end{aligned}
$$

But $tr\left(J\tilde{Y}D\tilde{Z}(\tilde{X})\right) = -tr\left(\tilde{Y}JD\tilde{Z}(\tilde{X})\right)$ (since $\tilde{Y}J = -J\tilde{Y}$), and since $tr(XY) = tr(YX)$ this in turn is equal to

$$
-tr\left(J\left(D\tilde{Z}(\tilde{X})\right)\tilde{Y}\right) \; .
$$

Consequently, the sum of the first and sixth terms in the expression above is zero. Similarly, the remaining terms cancel and $d\Omega(\tilde{X},\tilde{Y},\tilde{Z}) = 0$.

We summarize this: The Hermitian metric $\langle\cdot,\cdot\rangle$ on the Teichmüller space of a two dimensional surface M of genus greater than one, which is induced from the L^2-metric on the space of almost complex structures on M, is Kähler. Since $2\langle\cdot,\cdot\rangle_{WP} = \langle\cdot,\cdot\rangle$, we have

Theorem 5.1.10 *The Weil-Petersson metric is Kähler.*

5.2   The Natural Algebraic Connection on $\mathcal{A}$

Define the algebraic symmetric connection ∇ on $\mathcal{A}$ by

$$\nabla_Y X = DX(Y) - \frac{1}{2}J(XY + YX) \tag{5.6}$$

where D denotes derivative and where X and Y are vector fields on $\mathcal{A}$. One can easily show that $\nabla_Y X \in T_J \mathcal{A}$ if $X, Y \in T_J \mathcal{A}$. To see this one differentiates the relation $JX = -XJ$ in the direction Y obtaining the relation

$$JDX(Y) + YX = -XY - DX(Y)J \ .$$

Then

$$\begin{aligned}
J \cdot \nabla_Y X &= JDX(Y) + \tfrac{1}{2}(XY + YX) \\
&= -\tfrac{1}{2}(XY + YX) - DX(Y)J \\
&= -\nabla_Y X \cdot J \ .
\end{aligned}$$

An interesting relationship between the almost complex structure Φ and ∇ is given by

Theorem 5.2.1 *The covariant derivative of the almost complex structure Φ on $\mathcal{A}$ with respect to ∇ is zero; i.e. $\nabla\Phi \equiv 0$.*

PROOF: The 1-2 tensor $\nabla\Phi$ is defined by the first line of the following equation:

$$\begin{aligned}
(\nabla\Phi)(X, Y) &= \nabla_Y(\Phi X) - \Phi\nabla_Y X \\
&= \nabla_Y(JX) - J\nabla_Y X \\
&= D(JX)Y - \tfrac{1}{2}J(JXY + YJX) - J\left\{DX(Y) - \tfrac{1}{2}J(XY + YX)\right\} \\
&= YX + JDX(Y) + \tfrac{1}{2}(XY - YX) - JDX(Y) - \tfrac{1}{2}(XY + YX) \\
&= 0 \ .
\end{aligned}$$

$\blacksquare$

There are several ways to view this connection. First repeat the argument on page 85: if X and Y are vector fields on $\mathcal{A}$, then for each $J \in \mathcal{A}$, $X(J)$ and $Y(J)$ are trace free 1-1 tensors on M which are symmetric with respect to $g(J)$. Then $DY(X)(J)$ will be trace free but not symmetric. We defined the projection map Π by

$$\Pi(J) : Z \mapsto \frac{1}{2}(Z + Z^*)$$

where $*$ denotes the adjoint with respect to $\tilde{\Psi}(J)$. Then, differentiating the tangency condition $\Pi(J)X(J) = X(J)$ for X in direction $Y(J)$ and using (4.3), one immediately checks that

$$\nabla_X Y = DY(X) - D\Pi(X, Y) = \Pi DY(X) \ .$$

Given a metric $\langle\langle \cdot, \cdot \rangle\rangle$ on a Riemannian manifold P there is a unique symmetric connection $\hat{\nabla}$ associated to $\langle\langle \cdot, \cdot \rangle\rangle$ called the Levi-Cività connection. $\hat{\nabla}$ is a bilinear mapping on vector fields

$$\hat{\nabla} : \mathcal{X}(P) \times \mathcal{X}(P) \rightarrow \mathcal{X}(P)$$

uniquely characterized by the relations

$$X\langle\langle V, W \rangle\rangle = \langle\langle \hat{\nabla}_X V, W \rangle\rangle + \langle\langle V, \hat{\nabla}_X W \rangle\rangle$$

$$\hat{\nabla}_V W - \hat{\nabla}_W V = [V, W] \ , \tag{5.7}$$

where $[\cdot, \cdot]$ denotes the Lie bracket of vector fields,

$$[X, Y] = DY(X) - DX(Y) \ .$$

Indeed, the equations (5.7) imply

$$\begin{aligned} 2\langle\langle \nabla_Y X, W \rangle\rangle \ = \ & Y\langle\langle X, W \rangle\rangle + X\langle\langle Y, W \rangle\rangle - W\langle\langle X, Y \rangle\rangle \\ & + \langle\langle [Y, X], W \rangle\rangle - \langle\langle [Y, W], X \rangle\rangle - \langle\langle [X, W], Y \rangle\rangle \ . \end{aligned} \tag{5.8}$$

Remarkably ∇ is a *metric connection*. Let $g_0 \in \mathcal{M}$ be a fixed metric and define the Riemannian inner product $(\cdot, \cdot)$ on $\mathcal{A}$ by

$$(X, Y)_J = \int_M tr\,(X(J)Y(J))\,d\mu_{g_0} \ .$$

Note that this inner product differs from the one considered in section 2.5. It uses the *fixed* volume element $d\mu_{g_0}$ for integration. We leave it to the reader to verify that ∇ is the Levi-Cività connection of $(\cdot, \cdot)$.

Let $\hat{\nabla}$ now denote the Levi-Cività connection of the L^2-metric on $\mathcal{A}$. At first glance the algebraic connection ∇ should have nothing to do with $\hat{\nabla}$. Surprisingly, however the following result shows that they induce the same connection on the quotient space $\mathcal{T}(M) = \mathcal{A}/\mathcal{D}_0$.

Theorem 5.2.2 *If $\hat{\nabla}$ denotes the Levi-Cività connection of the L^2-metric $\langle\langle \cdot, \cdot \rangle\rangle$ on $\mathcal{A}$, then the horizontal components of $\nabla_Y X$ and $\hat{\nabla}_Y X$ agree if X and Y are horizontal.*

PROOF: To show this it suffices to check that equations (5.7) are verified for ∇ on horizontal fields $\tilde{X}, \tilde{Y}$ and $\tilde{W}$, or

$$\tilde{X} \left\langle\left\langle \tilde{V}, \tilde{W}\right\rangle\right\rangle = \left\langle\left\langle \nabla_{\tilde{X}} \tilde{V}, \tilde{W}\right\rangle\right\rangle + \left\langle\left\langle \tilde{V}, \nabla_{\tilde{X}} \tilde{W}\right\rangle\right\rangle$$

and

$$\nabla_{\tilde{V}} \tilde{W} - \nabla_{\tilde{W}} \tilde{V} = [\tilde{V}, \tilde{W}] \ .$$

This is indeed sufficient, because formula (5.8) is a consequence of formulas (5.7) by simple algebra, and so we need the latter only for horizontal vector fields in order to get the former for horizontal vector fields. Moreover, the vertical components of the covariant derivatives in (5.8) do not affect the value of the scalar products since the splitting in vertical and horizontal components is orthogonal.

The second equation follows at once from the definition of ∇. For the first, compute

$$\tilde{X} \left\langle\left\langle \tilde{V}, \tilde{W}\right\rangle\right\rangle_J = \tilde{X} \left\{ \int\limits_M tr(\tilde{V}\tilde{W}) d\mu_{g(J)} \right\} \ , \tag{5.9}$$

where we again denote $\tilde{\Psi}(J)$ by the more suggestive notation $g(J)$. Since $D\mu_{g(J)}(\tilde{X}) \equiv 0$ if $\tilde{X}$ is horizontal (cf. lemma 5.1.7), (5.9) is equal to

$$\int\limits_M tr\left(\left(D\tilde{V}(\tilde{X})\right) \tilde{W}\right) d\mu_{g(J)} + \int\limits_M tr\left(\tilde{V}\left(D\tilde{W}(\tilde{X})\right)\right) d\mu_{g(J)} \ . \tag{5.10}$$

We already know by lemma 5.1.8 that for *any* three tangent vectors $\tilde{X}, \tilde{Y}, \tilde{Z} \in T_J\mathcal{A}$ we have

$$tr(\tilde{X}\tilde{Y}\tilde{Z}) = 0 \ .$$

Similarly we see that $tr(J\tilde{X}\tilde{Y}\tilde{Z}) = 0$. From this it follows that (5.10) is equal to

$$\int\limits_M tr\left(\left(D\tilde{V}(\tilde{X}) - \tfrac{1}{2}J(\tilde{X}\tilde{V} + \tilde{V}\tilde{X})\right) \tilde{W}\right) d\mu_{g(J)}$$
$$+ \int\limits_M tr\left(\tilde{V}\left(D\tilde{W}(\tilde{X}) - \tfrac{1}{2}J(\tilde{W}\tilde{X} + \tilde{X}\tilde{W})\right)\right) d\mu_{g(J)}$$
$$= \left\langle\left\langle \nabla_{\tilde{X}} \tilde{V}, \tilde{W}\right\rangle\right\rangle_J + \left\langle\left\langle \tilde{V}, \nabla_{\tilde{X}} \tilde{W}\right\rangle\right\rangle_J$$

which concludes the proof of 5.2.2.   ■

Theorems 5.2.1 and 5.2.2 now permit us to give a very quick proof that the Weil-Petersson metric is Kähler with respect to the induced almost complex structures $\hat{\Phi}$ on $\mathcal{T}(M)$.

In view of theorem 5.2.2 denote by $\hat{\nabla}^H$ the Levi-Cività connection of the Weil-Petersson metric. Formally the superscript H means "horizontal component". An alternative, but equivalent definition of Kähler is that

$$\hat{\nabla}^H \hat{\Phi} \equiv 0 \ .$$

Theorem 5.2.3 *The Weil-Petersson metric is Kähler.*

SECOND PROOF: If X, Y are vector fields on $\mathcal{T}(M)$ and $\tilde{X}, \tilde{Y}$ their horizontal lifts to $\mathcal{A}$ then

$$(\hat{\nabla}^H\hat{\Phi})(X, Y) = D\pi\left\{\hat{\nabla}_{\tilde{X}}(\Phi\tilde{Y}) - \Phi\hat{\nabla}_{\tilde{X}}\tilde{Y}\right\} \; . \tag{5.11}$$

Since the almost complex structure Φ on $\mathcal{A}$ preserves horizontal and vertical vectors (and in fact preserves the L^2-splitting into horizontal and vertical), then using the last theorem the right hand side of (5.11) is equal to

$$D\pi\left\{\nabla_{\tilde{X}}\Phi\tilde{Y} - \Phi\nabla_{\tilde{X}}\tilde{Y}\right\} \equiv 0$$

by theorem 5.2.1.   ∎

We can also compute the curvature tensor $\mathcal{R}(X, Y)Z$ of ∇ which is defined by

$$(\nabla_X\nabla_Y - \nabla_Y\nabla_X - \nabla_{[X,Y]})Z = \mathcal{R}(X, Y)Z \; .$$

Now, since we know ∇ *explicitly* this curvature tensor is easy to compute, namely

$$\begin{aligned}
\nabla_X\nabla_Y Z &= D\left\{DZ(Y) - \tfrac{1}{2}(JZY + JYZ)\right\}(X) \\
&\quad - \tfrac{1}{2}J\left\{X\left[DZ(Y) - \tfrac{1}{2}(JZY + JYZ)\right] + \left[DZ(Y) - \tfrac{1}{2}(JZY + JYZ)\right]X\right\} \\
&= D^2Z(X, Y) + DZDY(X) - \tfrac{1}{2}XZY - \tfrac{1}{2}JDZ(X)Y - \tfrac{1}{2}JZDY(X) \\
&\quad - \tfrac{1}{2}XYZ - \tfrac{1}{2}JDY(X)Z - \tfrac{1}{2}JYDZ(X) - \tfrac{1}{2}JXDZ(Y) \\
&\quad + \tfrac{1}{4}XZY + \tfrac{1}{4}XYZ - \tfrac{J}{2}DZ(Y)X - \tfrac{1}{4}ZYX - \tfrac{1}{4}YZX \; .
\end{aligned}$$

$\nabla_Y\nabla_X Z$ is obtained by interchanging X and Y in the last computation. Hence we find that

$$\mathcal{R}(X, Y)Z = \frac{1}{4}(-XYZ - ZYX + YXZ + ZXY) \; .$$

Therefore

$$\langle\langle\mathcal{R}(X, Y)Y, X\rangle\rangle_J = \frac{1}{2}\int_M trace(-Y^2X^2 + YXYX)d\mu_{g(J)} \; . \tag{5.12}$$

Thus for fixed $J \in \mathcal{A}$ let $X(J), Y(J) \in T_J\mathcal{A}$. Furthermore for each $x \in M$ let us denote the matrices of $X(J)_x$ and $Y(J)_x$ by $\left(\begin{smallmatrix} c & d \\ d & -c \end{smallmatrix}\right)$ and $\left(\begin{smallmatrix} a & b \\ b & -a \end{smallmatrix}\right)$. Then

$$trace(-Y^2X^2 + YXYX) = -4(ad - bc)^2 < 0 \tag{5.13}$$

for linearly independent X and Y, and this holds whether or not X or Y is horizontal.

This suggests (but, by no means proves) that the sectional curvature of Teichmüller space with respect to the Weil-Petersson metric might be negative. The proof of this fact will constitute section 4 of this chapter. Before providing a proof, we must investigate the properties of our new connection.

5.3   Further Properties of the Algebraic Connection and the non-Integrability of the Horizontal Distribution on $\mathcal{A}$

In this section we choose to use the notation $D_Y X$ instead of $DX(Y)$ in order to maintain the analogy with the notation $\nabla_Y X$. Moreover it will be understood that δ_g means $\delta_{g(J)}$ thus avoiding clumsy notation. The main result of this section is:

Theorem 5.3.1 *If $\tilde{X}$ and $\tilde{Y}$ are horizontal, then*

$$\delta_g[\tilde{Y}, \tilde{X}] = \delta_g(\nabla_{\tilde{Y}}\tilde{X} - \nabla_{\tilde{X}}\tilde{Y}) = \delta_g(D_{\tilde{Y}}\tilde{X} - D_{\tilde{X}}\tilde{Y}) = d\lambda \qquad (5.14)$$

$$\delta_g(D_{\tilde{X}}\tilde{Y} + D_{\tilde{Y}}\tilde{X}) = -*d\mu \qquad (5.15)$$

$$\delta_g(\nabla_{\tilde{X}}\tilde{Y} + \nabla_{\tilde{Y}}\tilde{X}) = +*d\mu \qquad (5.16)$$

where $\lambda, \mu : M \to \mathbb{R}$ are the functions

$$\begin{aligned} \lambda(x) &= (ad - bc)(x) &= \tfrac{1}{4}tr\left(J(\tilde{X}\tilde{Y} - \tilde{Y}\tilde{X})\right) \\ and \quad \mu(x) &= (ac + bd)(x) &= \tfrac{1}{4}tr(\tilde{X}\tilde{Y} + \tilde{Y}\tilde{X}) \end{aligned}$$

and if $w = \zeta dx + \eta dy$ in conformal coordinates, the Hodge dual $*$ is given by:

$$*w = -\eta dx + \zeta dy \ .$$

These formulas will follow directly from the next

Lemma 5.3.2 *Let $\tilde{X}, \tilde{Y}$ be the horizontal 1-1 tensors which in local coordinates are representable by the matrices $\begin{pmatrix} a & b \\ b & -a \end{pmatrix}$ and $\begin{pmatrix} c & d \\ d & -c \end{pmatrix}$, respectively. Taking the divergence of a 1-1 tensor in conformal coordinates $g_{ij} = p\delta_{ij}$ on M, the divergence of the derivative of $\tilde{X}$ in the direction $\tilde{Y}$ is given by*

$$\begin{aligned} \delta_g D_{\tilde{Y}}\tilde{X} &= \left(-\tfrac{1}{p}\tfrac{\partial p}{\partial y}(ac + bd) + a\tfrac{\partial d}{\partial x} - b\tfrac{\partial c}{\partial x}\right) dx \\ &+ \left(+\tfrac{1}{p}\tfrac{\partial p}{\partial x}(ac + bd) + a\tfrac{\partial d}{\partial y} - b\tfrac{\partial c}{\partial y}\right) dy \end{aligned}$$

where we write the coordinate (x^1, x^2) as (x, y). Here we take the divergence of a 1-1 tensor to be a 1-form.

PROOF: The divergence of the tensor a_i^j is given by the expression

$$a_{i;j}^j = \frac{1}{\sqrt{g}}\frac{\partial}{\partial x^j}(a_i^j \sqrt{g}) - a_\ell^j \Gamma_{ij}^\ell$$

where $\Gamma^\ell_{ij} = \frac{1}{2}g^{\ell k}\left\{\frac{\partial g_{ik}}{\partial x^j} + \frac{\partial g_{jk}}{\partial x^i} - \frac{\partial g_{ij}}{\partial x^k}\right\}$.

Here, a^j_i are functions of both $J \in \mathcal{A}$ and $x \in M$. The differentiations D acting on J and δ acting on x commute of course. Moreover since $\tilde{Y}$ is trace free, we may treat $\sqrt{g}$ like a constant with respect to the differentiation D.

Clearly if $\tilde{X}$ and $\tilde{Y}$ are horizontal (and therefore divergence free) we see that

$$(\delta_g D_{\tilde{Y}}\tilde{X})_i = a^j_\ell D\Gamma^\ell_{ij}(h) \ , \tag{5.17}$$

where D denotes the derivative of Γ in the direction of the symmetric tensor $h = D\left(g(J)\right)\tilde{Y}$ where $D\left(g(J)\right) = D\tilde{\Psi}(J)$ represents the derivative of the map $J \mapsto g(J) = \tilde{\Psi}(J)$.

Now a straightforward computation shows that

$$D\Gamma^\ell_{ij}(h) = -\frac{1}{2}h^{\ell i}\frac{\partial p}{\partial x^j} - \frac{1}{2}h^{\ell j}\frac{\partial p}{\partial x^i} + \frac{1}{2}h^{\ell k}\frac{\partial p}{\partial x^k}\delta_{ij} + \frac{1}{2p}\left\{\frac{\partial h_{i\ell}}{\partial x^j} + \frac{\partial h_{j\ell}}{\partial x^i} - \frac{\partial h_{ij}}{\partial x^\ell}\right\} \ .$$

Since a^i_j, h_{ij} are symmetric, $a^j_\ell\frac{\partial h_{i\ell}}{\partial x^j} = a^j_\ell\frac{\partial h_{ij}}{\partial x^\ell}$, and so

$$a^j_\ell D\Gamma^\ell_{ij}(h) = -\frac{1}{2}a^j_\ell h^{\ell i}\frac{\partial p}{\partial x^j} - \frac{1}{2}a^j_\ell h^{\ell j}\frac{\partial p}{\partial x^i} + \frac{1}{2}a^j_\ell h^{\ell k}\frac{\partial p}{\partial x^k}\delta_{ij} + \frac{a^j_\ell}{2p}\left\{\frac{\partial h_{j\ell}}{\partial x^i}\right\} \ .$$

If H is a trace free divergence free 1-1 tensor then by formula (2.6)

$$Dg(J)(H) = (-JH)_\flat = (HJ)_\flat \ .$$

Using this fact for $H = \tilde{Y}$, we see that $((h^{ij})) = \frac{1}{p}\begin{pmatrix} d & -c \\ -c & -d \end{pmatrix}$, and for $i = 1$

$$-\frac{1}{2}a^j_\ell h^{\ell i}\frac{\partial p}{\partial x^j} = -\frac{1}{2p}\left\{\frac{\partial p}{\partial x}(ad - bc) + \frac{\partial p}{\partial y}(ac + bd)\right\}$$

and for $i = 2$, this is equal to

$$-\frac{1}{2p}\left\{\frac{\partial p}{\partial x}(-ac - bd) + \frac{\partial p}{\partial y}(-bc + ad)\right\} \ .$$

Moreover the term

$$-\frac{1}{2}a^j_\ell h^{\ell j}\frac{\partial p}{\partial x^i} = -\frac{1}{p}\frac{\partial p}{\partial x^i}(ad - bc)$$

and for $i = 1$

$$+\frac{1}{2}a^j_\ell h^{\ell k}\frac{\partial p}{\partial x^k}\delta_{ij} = +\frac{1}{2p}\left\{\frac{\partial p}{\partial x}(ad - bc) + \frac{\partial p}{\partial y}(-ac - bd)\right\}$$

and for $i = 2$ this is equal to

$$+ \frac{1}{2p} \left\{ \frac{\partial p}{\partial x}(ac + bd) + \frac{\partial p}{\partial y}(-bc + ad) \right\} \quad .$$

Finally

$$\frac{1}{2p} a_\ell^j \frac{\partial h_{j\ell}}{\partial x^i} = \frac{1}{2p} a_\ell^j \frac{\partial}{\partial x^i}(ph_j^\ell) = +\frac{1}{p} \left\{ a\frac{\partial}{\partial x^i}(pd) - b\frac{\partial}{\partial x^i}(pc) \right\} \quad .$$

Adding all terms together we obtain the conclusion of lemma 5.3.2.   ■

As a result we are now ready to complete the proof of our main theorem 5.3.1.

PROOF (5.3.1):   The formula for $\delta_g[\tilde{Y}, \tilde{X}]$ follows immediately from lemma 5.3.2. For the second formula (5.15) we use the fact that

$$-a\frac{\partial d}{\partial x} + b\frac{\partial c}{\partial x} + d\frac{\partial a}{\partial x} - c\frac{\partial b}{\partial x} = -\frac{a}{p}\frac{\partial}{\partial x}(pd) + \frac{d}{p}\frac{\partial}{\partial x}(pa) - \frac{c}{p}\frac{\partial}{\partial x}(pb) + \frac{b}{p}\frac{\partial}{\partial x}(pc)$$

and since $\tilde{X}$ and $\tilde{Y}$ are divergence free this is equal to

$$-\left\{ \frac{a}{p}\frac{\partial}{\partial y}(pc) + \frac{d}{p}\frac{\partial}{\partial y}(pb) + \frac{c}{p}\frac{\partial}{\partial y}(pa) + \frac{b}{p}\frac{\partial}{\partial y}(pd) \right\}$$

and similarly for x replaced by y. The second result then follows immediately.

The third formula (5.16) follows from the fact that

$$\delta_g J(\tilde{X}\tilde{Y} + \tilde{Y}\tilde{X}) = -2 *d\mu .$$

This concludes the proof of theorem 5.3.1.   ■

As a consequence of these computations we obtain a second corollary, namely the non-integrability of the horizontal distribution on $\mathcal{A}$.

Theorem 5.3.3 *The horizontal distribution on the bundle $\mathcal{A}$ is non-integrable.*

PROOF: The distribution is integrable iff

$$[\tilde{Y}, \tilde{X}]^\nu = 0$$

for all $\tilde{X}, \tilde{Y}$. See e.g. [103], vol. 1, ch. 6 for this fact. Fix $J \in \mathcal{A}$ arbitrarily. By theorem 5.3.1, $\delta_g[\tilde{Y}, \tilde{X}]^v = d\lambda$. Thus $[\tilde{Y}, \tilde{X}]^v = 0$ implies that $d\lambda \equiv 0$, i.e. λ is a constant function on M. For horizontal $\tilde{X}, \tilde{Y}$, in conformal coordinates $g_{ij} = p\delta_{ij}$, $(pa - ipb)dz^2$ and $(pc - ipd)dz^2$ are holomorphic quadratic differentials on M. Since every holomorphic quadratic differential on a surface of genus q has at least $4q - 4$ zeros (see [104]), there exists a point on M, where $a = b = 0$, i.e. where λ vanishes. Therefore $\lambda \equiv 0$. This argument holds for every J. But $\lambda \equiv 0$ identically on $\mathcal{A}$ immediately implies that $\tilde{X}$ and $\tilde{Y}$ are linearly dependent. Taking $\tilde{Y} = J\tilde{X}$ for some non-zero $\tilde{X}$, we see that $\tilde{X} = 0$. ∎

Although $[\tilde{X}, \tilde{Y}]$ is not horizontal, the vertical component, $[\tilde{X}, \tilde{Y}]^v$ cannot be arbitrary as the following theorem shows.

Theorem 5.3.4 *Let $\tilde{X}, \tilde{Y}$ be horizontal on $\mathcal{A}$. Then*

$$[\tilde{X}, \tilde{Y}]^v = L_\beta J$$

where the divergence $\delta_g \beta = 0$. Thus the vertical part of the Lie bracket must be tangent to the action of the group of volume preserving diffeomorphisms. Here $\delta_g \beta$ is given in local coordinates by

$$\delta_g \beta = \frac{1}{\sqrt{g}} \frac{\partial}{\partial x^i} (\beta^i \sqrt{g}) \ .$$

PROOF: Consider again the map $J \mapsto \tilde{\Psi}(J)$ and the derivative $H \mapsto D\tilde{\Psi}(J)H$. If $H = [\tilde{X}, \tilde{Y}]^v$, let $h = (HJ)_\flat$. Then from formula (2.6) it follows that

$$D\tilde{\Psi}(J)H = h + \rho g$$

where

$$\Delta\rho - \rho = \delta_g \delta_g h \ ,$$

Δ the Laplace-Beltrami operator and $\delta_g \delta_g$ the double covariant divergence of h with respect to the Poincaré metric $g(J) = \tilde{\Psi}(J)$.

But $HJ = -JH = J[\tilde{Y}, \tilde{X}]^v$.

Since $\delta_g[\tilde{Y}, \tilde{X}] = d\lambda$, and noting that if $\delta_g K = d\lambda$, then $\delta_g(JK) = *d\lambda$, we see that

$$\delta_g \delta_g h = \delta_g *d\lambda = -dd\lambda = 0 \ .$$

Therefore $\delta_g \delta_g h = 0$ which implies that $\rho = 0$. Since $\tilde{\Psi} : \mathcal{A} \to \mathcal{M}_{-1}$ is $\mathcal{D}$-equivariant, for $f \in \mathcal{D}$

$$\tilde{\Psi}(f^*J) = f^*\tilde{\Psi}(J) \ .$$

From this it follows that the differential of $\tilde{\Psi}$ takes vertical to vertical and for $g = \tilde{\Psi}(J)$,

$$D\tilde{\Psi}(J)L_\beta J = L_\beta g(J) = L_\beta g = (HJ)_\flat = h \; .$$

However,

$$\delta_g \beta = \frac{1}{2} tr_g \{L_\beta g\} = \frac{1}{2} tr_g h \equiv 0 \; ,$$

concluding the proof of theorem 5.3.4.   ∎

We now continue to investigate the properties of ∇ on the bundle $\mathcal{A}$. We have already remarked that the horizontal component of our algebraic connection agrees with the Levi-Cività connection on horizontal vector fields on $\mathcal{A}$. However, as we shall presently see, these connections agree on a larger class of vector fields.

Theorem 5.3.5 *Let V and Z represent horizontal vector fields on $\mathcal{A}$, and $W = L_\beta J$, a vertical field on $\mathcal{A}$.*

$$\begin{aligned}
2\langle\langle \nabla_V Z, W \rangle\rangle_J = \; & V\langle\langle Z, W \rangle\rangle_J + Z\langle\langle V, W \rangle\rangle_J - W\langle\langle V, Z \rangle\rangle_J \\
& + \langle\langle [V, Z], W \rangle\rangle_J - \langle\langle [V, W], Z \rangle\rangle_J - \langle\langle [Z, W], V \rangle\rangle_J \\
& + \int tr(ZV)(\delta_g\beta)d\mu_{g(J)} \; .
\end{aligned}$$

PROOF: First we remark the distinction, seen in this formula, between the Levi-Cività connection $\hat{\nabla}$ and the connection ∇. For $\hat{\nabla}$ the last term would not be present.

From the computation in theorem 5.2.2 we see that

$$V\langle\langle Z, W \rangle\rangle = \langle\langle \nabla_V Z, W \rangle\rangle + \langle\langle Z, \nabla_V W \rangle\rangle \tag{5.18}$$

$$Z\langle\langle V, W \rangle\rangle = \langle\langle \nabla_Z V, W \rangle\rangle + \langle\langle V, \nabla_Z W \rangle\rangle \; . \tag{5.19}$$

But since W is not horizontal we see that

$$W\langle\langle V, Z \rangle\rangle = \langle\langle \nabla_W V, Z \rangle\rangle + \langle\langle V, \nabla_W Z \rangle\rangle + \int_M tr(ZV)D_W\{d\mu_{g(J)}\} \; . \tag{5.20}$$

Consider the map $g \mapsto \mu_g$. In coordinates this map is given by $g \mapsto \sqrt{g}dx\,dy$. Thus the derivative of this map is the linear map

$$h \mapsto \left(\frac{1}{2} tr_g h\right)\mu_g \; ,$$

$tr_g h$ the trace of h with respect to g (which of course vanishes if h is trace free). Since, as we have just seen, $D\check\Psi$ takes vertical to vertical, using the chain rule we have

$$D_W\left(d\mu_{g(J)}\right) = \frac{1}{2}tr_g(L_\beta g)\cdot d\mu_{g(J)} = \delta_g\beta\cdot d\mu_{g(J)} \ . \tag{5.21}$$

Subtracting (5.20) from the sum of (5.18) and (5.19) and applying (5.21) we have the proof of theorem 5.3.5. ∎

As a corollary to this proof we have the following result.

Corollary 5.3.6 *If V, W and Z are either horizontal fields on $\mathcal{A}$, or vertical fields which are tangent to the action of the group of volume preserving diffeomorphisms on $\mathcal{A}$, then*

$$\langle\langle\nabla_V Z, W\rangle\rangle = \langle\langle\hat\nabla_V Z, W\rangle\rangle \ ,$$

where $\hat\nabla$ is the Levi-Cività connection of $\langle\langle\cdot,\cdot\rangle\rangle$.

In the next section we use these calculations to compute the sectional curvature of the Weil-Petersson metric on $\mathcal{T}(M)$.

5.4 The Curvature of Teichmüller Space with Respect to its Weil-Petersson Metric

In this section we shall derive a formula for the sectional curvature of Teichmüller space with respect to its Weil-Petersson metric. As a consequence of this formula we will show that the sectional curvature is negative and moreover there are explicit bounds on the Ricci and holomorphic sectional curvatures.

In [5] Ahlfors showed that the holomorphic sectional curvature of the Weil-Petersson metric is negative and that it is Ricci negative. Tromba [112] first computed this curvature and somewhat later Wolpert [125] also derived this result. Since then the curvature was also computed by Siu [99] and Jost [57].

The Riemann curvature tensor $\hat{\mathcal{R}}$ of the metric $\langle\langle\cdot,\cdot\rangle\rangle$ is defined in terms of the associated Levi-Cività connection $\hat\nabla$

$$\left\{\hat\nabla_X\hat\nabla_Y - \hat\nabla_Y\hat\nabla_X - \hat\nabla_{[X,Y]}\right\} Z = \hat{\mathcal{R}}(X,Y)Z \ . \tag{5.22}$$

$\hat{\mathcal{R}}$ will be distinguished from $\mathcal{R}$, which is defined by the same formula as $\hat{\mathcal{R}}$ with the algebraic connection ∇ instead of $\hat{\nabla}$. $\hat{\mathcal{R}}$ is a trilinear mapping on $\mathcal{X}(\mathcal{A})$. The sectional curvature $\mathcal{K}$ at a point $J \in \mathcal{A}$ in orthonormal planar directions $X(J)$, $Y(J) \in T_J\mathcal{A}$ is given by

$$\mathcal{K}(X,Y)_J = \langle\langle \hat{\mathcal{R}}(X,Y)Y, X\rangle\rangle_J \ .$$

We have no need to write $\hat{\mathcal{K}}$ instead of $\mathcal{K}$ since no corresponding expression will be used for the algebraic connection. Since $\pi : \mathcal{A} \to \mathcal{A}/\mathcal{D}_0$ is a Riemannian submersion with respect to the L^2-metric on $\mathcal{A}$ and the induced metric on $\mathcal{T}(M)$, one would like to apply Barrett O'Neill's curvature formula [87] for a Riemannian submersion to compute the induced curvature on $\mathcal{T}(M)$. For this, however, it would be necessary to compute the Levi-Cività connection of the L^2-metric on $\mathcal{A}$, which appears to be rather difficult.

Our trick to compute this curvature will be to use the algebraic connection we introduced in section 5.2.

Let $\mathcal{K}$ denote the curvature of Teichmüller space $\mathcal{T}(M) = \mathcal{A}/\mathcal{D}_0$ with respect to its Weil-Petersson metric. If X and Y now denote orthonormal vector fields on $\mathcal{T}(M)$ let $\tilde{X}$ and $\tilde{Y}$ denote the unique horizontal lifts with respect to the L^2-metric.

Since the metric $\langle\langle \cdot, \cdot \rangle\rangle$ induces on the quotient space twice the Weil-Petersson metric (cf. theorem 2.6.1) we see that if $\hat{\nabla}$ denotes the Levi-Cività connection on $\mathcal{A}$,

$$2 \cdot \mathcal{K}(X,Y) = \left\langle\left\langle \hat{\nabla}_{\tilde{X}}(\hat{\nabla}_{\tilde{Y}}\tilde{Y})^H - \hat{\nabla}_{\tilde{Y}}(\hat{\nabla}_{\tilde{X}}\tilde{Y})^H - \hat{\nabla}_{[\tilde{X},\tilde{Y}]^H}\tilde{Y}, \tilde{X} \right\rangle\right\rangle \qquad (5.23)$$

where the superscripts H and V denote horizontal and vertical components, respectively.

Since $(\hat{\nabla}_{\tilde{X}}\tilde{Y})^H = (\nabla_{\tilde{X}}\tilde{Y})^H$ we see that

$$
\begin{aligned}
2\mathcal{K}(X,Y) &= \left\langle\left\langle \nabla_{\tilde{X}}(\nabla_{\tilde{Y}}\tilde{Y})^H - \nabla_{\tilde{Y}}(\nabla_{\tilde{X}}\tilde{Y})^H - \nabla_{[\tilde{X},\tilde{Y}]^H}\tilde{Y}, \tilde{X} \right\rangle\right\rangle \qquad (5.24) \\
&= \left\langle\left\langle \nabla_{\tilde{X}}\nabla_{\tilde{Y}}\tilde{Y} - \nabla_{\tilde{Y}}\nabla_{\tilde{X}}\tilde{Y} - \nabla_{[\tilde{X},\tilde{Y}]}\tilde{Y}, \tilde{X} \right\rangle\right\rangle \\
&\quad - \left\langle\left\langle \nabla_{\tilde{X}}(\nabla_{\tilde{Y}}\tilde{Y})^V, \tilde{X} \right\rangle\right\rangle + \left\langle\left\langle \nabla_{\tilde{Y}}(\nabla_{\tilde{X}}\tilde{Y})^V, \tilde{X} \right\rangle\right\rangle + \left\langle\left\langle \nabla_{[\tilde{X},\tilde{Y}]^V}\tilde{Y}, \tilde{X} \right\rangle\right\rangle \\
&= \left\langle\left\langle \mathcal{R}(\tilde{X},\tilde{Y})\tilde{Y}, \tilde{X} \right\rangle\right\rangle + \left\langle\left\langle (\nabla_{\tilde{Y}}\tilde{Y})^V, \nabla_{\tilde{X}}\tilde{X} \right\rangle\right\rangle \\
&\quad - \left\langle\left\langle (\nabla_{\tilde{X}}\tilde{Y})^V, \nabla_{\tilde{Y}}\tilde{X} \right\rangle\right\rangle + \left\langle\left\langle \nabla_{[\tilde{X},\tilde{Y}]^V}\tilde{Y}, \tilde{X} \right\rangle\right\rangle \ .
\end{aligned}
$$

Our goal will now be to compute explicitly the last three terms in the expression immediately above. We begin with

Lemma 5.4.1 $\left\langle\left\langle \nabla_{[\tilde{X},\tilde{Y}]^V}\tilde{Y}, \tilde{X} \right\rangle\right\rangle = -\left\langle\left\langle [\tilde{X},\tilde{Y}]^V, \nabla_{\tilde{Y}}\tilde{X} \right\rangle\right\rangle.$

PROOF:

$$\left\langle\!\left\langle \nabla_{[\tilde{X},\tilde{Y}]^v}\tilde{Y},\tilde{X}\right\rangle\!\right\rangle = \left\langle\!\left\langle \nabla_{\tilde{Y}}[\tilde{X},\tilde{Y}]^v,\tilde{X}\right\rangle\!\right\rangle - \left\langle\!\left\langle \left[\tilde{Y},[\tilde{X},\tilde{Y}]^v\right],\tilde{X}\right\rangle\!\right\rangle \ . \qquad (5.25)$$

If $\pi : \mathcal{A} \to \mathcal{T}(M)$ and $\langle\cdot,\cdot\rangle$ denotes the induced metric on $\mathcal{T}(M) = \mathcal{A}/\mathcal{D}_0$ then

$$\langle X,Y\rangle_{[J]} = \langle\!\langle \tilde{X},\tilde{Y}\rangle\!\rangle_J \text{ where } D\pi(\tilde{X}) = X, \ D\pi(\tilde{Y}) = Y \ .$$

Moreover $\left[D\pi(J)\tilde{Y}, D\pi(J)\tilde{Z}\right] = D\pi(J)[\tilde{Y},\tilde{Z}]$. Considering the second term on the right hand side of the equality (5.25) we see that

$$D\pi\left[\tilde{Y},[\tilde{X},\tilde{Y}]^v\right] = \left[D\pi(\tilde{Y}), D\pi[\tilde{X},\tilde{Y}]^v\right] = 0$$

since $[\tilde{X},\tilde{Y}]^v \in \text{Ker } D\pi$. Thus

$$\left\langle\!\left\langle \left[\tilde{Y},[\tilde{X},\tilde{Y}]^v\right],\tilde{X}\right\rangle\!\right\rangle_J = \left\langle D\pi\left[\tilde{Y},[\tilde{X},\tilde{Y}]^v\right], D\pi(\tilde{X})\right\rangle_{[J]} = 0$$

whence

$$\left\langle\!\left\langle \nabla_{[\tilde{X},\tilde{Y}]^v}\tilde{Y},\tilde{X}\right\rangle\!\right\rangle = \left\langle\!\left\langle \nabla_{\tilde{Y}}[\tilde{X},\tilde{Y}]^v,\tilde{X}\right\rangle\!\right\rangle \ .$$

But

$$\left\langle\!\left\langle [\tilde{X},\tilde{Y}]^v,\tilde{X}\right\rangle\!\right\rangle \equiv 0$$

since $\tilde{X}$ is horizontal and $[\tilde{X},\tilde{Y}]^v$ is vertical. Thus

$$0 = \tilde{Y}\left\langle\!\left\langle [\tilde{X},\tilde{Y}]^v,\tilde{X}\right\rangle\!\right\rangle = \left\langle\!\left\langle \nabla_{\tilde{Y}}[\tilde{X},\tilde{Y}]^v,\tilde{X}\right\rangle\!\right\rangle + \left\langle\!\left\langle [\tilde{X},\tilde{Y}]^v,\nabla_{\tilde{Y}}\tilde{X}\right\rangle\!\right\rangle$$

from which we conclude lemma 5.4.1. ∎

Thus we arrive at a new formula for the curvature of Teichmüller space, with respect to the Weil-Petersson metric, namely

$$\begin{aligned}
2 \cdot \mathcal{K}(X,Y) &= \left\langle\!\left\langle \mathcal{R}(\tilde{X},\tilde{Y})\tilde{Y},\tilde{X}\right\rangle\!\right\rangle \qquad\qquad\qquad\qquad (5.26)\\
&+ \left\langle\!\left\langle (\nabla_{\tilde{Y}}\tilde{Y})^v,\nabla_{\tilde{X}}\tilde{X}\right\rangle\!\right\rangle - \left\langle\!\left\langle (\nabla_{\tilde{X}}\tilde{Y})^v,\nabla_{\tilde{Y}}\tilde{X}\right\rangle\!\right\rangle - \left\langle\!\left\langle [\tilde{X},\tilde{Y}]^v,\nabla_{\tilde{Y}}\tilde{X}\right\rangle\!\right\rangle \ .
\end{aligned}$$

We continue to calculate these terms explicitly.

Lemma 5.4.2 *Let T be a vector field on $\mathcal{A}$ tangent to the action of the group of volume preserving diffeomorphisms. Then if $\tilde{X},\tilde{Y}$ are horizontal lifts of vector fields X,Y on $\mathcal{T}(M)$,*

$$\left\langle\!\left\langle \nabla_{\tilde{Y}}\tilde{X},T\right\rangle\!\right\rangle = \frac{1}{2}\left\langle\!\left\langle [\tilde{Y},\tilde{X}],T\right\rangle\!\right\rangle \ .$$

PROOF: By theorem 5.3.5 we have that

$$2\left\langle\left\langle\nabla_{\tilde{Y}}\tilde{X},T\right\rangle\right\rangle \; = \; \tilde{Y}\left\langle\left\langle\tilde{X},T\right\rangle\right\rangle + \tilde{X}\left\langle\left\langle\tilde{Y},T\right\rangle\right\rangle - T\left\langle\left\langle\tilde{X},\tilde{Y}\right\rangle\right\rangle \tag{5.27}$$
$$+ \left\langle\left\langle[\tilde{Y},\tilde{X}],T\right\rangle\right\rangle - \left\langle\left\langle[\tilde{Y},T],\tilde{X}\right\rangle\right\rangle - \left\langle\left\langle[\tilde{X},T],\tilde{Y}\right\rangle\right\rangle \; .$$

Recall that we are denoting the induced metric by $\langle\cdot,\cdot\rangle$. Since $\tilde{X},\tilde{Y}$ are horizontal lifts and $\langle X,Y\rangle = \langle\langle\tilde{X},\tilde{Y}\rangle\rangle$ and T is vertical, $T\langle\langle\tilde{X},\tilde{Y}\rangle\rangle = 0$. The first two terms in (5.27) clearly vanish. Since $\tilde{X},\tilde{Y}$ are horizontal lifts and $[\tilde{X},T]$, $[\tilde{Y},T]$ are vertical, the last two vanish as in lemma 5.4.1 and this proves the lemma.   ■

Using theorem 5.3.4 that $[\tilde{X},\tilde{Y}]^v$ is tangent to the group of volume preserving diffeomorphisms we obtain:

Lemma 5.4.3

$$-\left\langle\left\langle[\tilde{X},\tilde{Y}]^v,\nabla_{\tilde{Y}}\tilde{X}\right\rangle\right\rangle = \frac{1}{2}\left\|[\tilde{X},\tilde{Y}]^v\right\|^2 \; .$$

Note that this is the last term in the formula (5.26) for the sectional curvature. Returning to this formula we now have

$$2\cdot\mathcal{K}(X,Y) \; = \; \left\langle\left\langle\mathcal{R}(\tilde{X},\tilde{Y})\tilde{Y},\tilde{X}\right\rangle\right\rangle + \left\langle\left\langle(\nabla_{\tilde{Y}}\tilde{Y})^v,\nabla_{\tilde{X}}\tilde{X}\right\rangle\right\rangle \tag{5.28}$$
$$- \left\langle\left\langle(\nabla_{\tilde{X}}\tilde{Y})^v,\nabla_{\tilde{Y}}\tilde{X}\right\rangle\right\rangle + \frac{1}{2}\left\|[\tilde{X},\tilde{Y}]^v\right\|^2 \; .$$

Now

$$-\left\langle\left\langle(\nabla_{\tilde{X}}\tilde{Y})^v,\nabla_{\tilde{Y}}\tilde{X}\right\rangle\right\rangle \; = \; -\left\langle\left\langle\nabla_{\tilde{X}}\tilde{Y},(\nabla_{\tilde{Y}}\tilde{X})^v\right\rangle\right\rangle$$
$$= \; -\left\langle\left\langle\nabla_{\tilde{X}}\tilde{Y},(\nabla_{\tilde{Y}}\tilde{X})^v\right\rangle\right\rangle + \left\langle\left\langle\nabla_{\tilde{Y}}\tilde{X},(\nabla_{\tilde{Y}}\tilde{X})^v\right\rangle\right\rangle$$
$$- \left\langle\left\langle\nabla_{\tilde{Y}}\tilde{X},(\nabla_{\tilde{Y}}\tilde{X})^v\right\rangle\right\rangle$$
$$= \; \left\langle\left\langle[\tilde{Y},\tilde{X}]^v,\nabla_{\tilde{Y}}\tilde{X}\right\rangle\right\rangle - \left\|(\nabla_{\tilde{Y}}\tilde{X})^v\right\|^2 \; .$$

Applying lemma 5.4.3 we see that this is equal to

$$\frac{1}{2}\left\|[\tilde{X},\tilde{Y}]^v\right\|^2 - \left\|(\nabla_{\tilde{Y}}\tilde{X})^v\right\|^2 \; .$$

Thus we arrive at our final formula for the curvature, namely

$$2\cdot\mathcal{K}(X,Y) \; = \; \left\langle\left\langle\mathcal{R}(\tilde{X},\tilde{Y})\tilde{Y},\tilde{X}\right\rangle\right\rangle + \left\langle\left\langle(\nabla_{\tilde{Y}}\tilde{Y})^v,(\nabla_{\tilde{X}}\tilde{X})^v\right\rangle\right\rangle \tag{5.29}$$
$$- \left\|(\nabla_{\tilde{Y}}\tilde{X})^v\right\|^2 + \left\|[\tilde{Y},\tilde{X}]^v\right\|^2 \; .$$

We proceed to evaluate each of these terms explicitly.

For example, we can now calculate the last term of formula (5.29):

Theorem 5.4.4 *Let X and Y be vector fields on Teichmüller's space $\mathcal{A}/\mathcal{D}_0$ and denote by $\tilde{X}, \tilde{Y}$ their horizontal lifts. Represent $\tilde{X}, \tilde{Y}$ in conformal coordinates by the matrices $\begin{pmatrix} a & b \\ b & -a \end{pmatrix}$ and $\begin{pmatrix} c & d \\ d & -c \end{pmatrix}$ respectively and let*

$$\lambda = ad - bc = \frac{1}{4} trace \left(J(\tilde{X}\tilde{Y} - \tilde{Y}\tilde{X}) \right) \ .$$

Then

$$\frac{1}{2} \cdot \left\| [\tilde{Y}, \tilde{X}]^v \right\|^2 = \int_M \lambda^2 d\mu_{g(J)} + \int_M (\mathcal{L}^{-1}\lambda)\lambda d\mu_{g(J)}$$

where $\mathcal{L}$ is the invertible elliptic operator on functions ρ given by

$$\mathcal{L}\rho = \Delta\rho - \rho \ .$$

Before proceeding with the proof we shall need:

Lemma 5.4.5 *Suppose $H \in T_J\mathcal{A}$ is vertical, $H = L_\beta J$ with $\delta_g H = *d\lambda$, for some smooth function $\lambda : M \to \mathbb{R}$. Then the divergence of β, $\delta_g\beta = \rho$ where*

$$\Delta\rho - \rho = \Delta\lambda \ .$$

PROOF: The proof is analogous to that of theorem 5.3.4. Let $h = (HJ)_\flat$. Then

$$D\tilde{\Psi}(J)H = h + \rho g = L_\beta g$$

$$\Delta\rho - \rho = \delta_g\delta_g h \ .$$

But $\delta_g h = -\delta_g(JH)_\flat = -*\delta_g H = -*(*d\lambda) = +d\lambda$. Therefore $\delta_g\delta_g h = \Delta\lambda$. But

$$\rho = \frac{1}{2}tr(h + \rho g) = \frac{1}{2}tr(L_\beta g) = \delta_g\beta \ .$$

∎

We proceed with the proof of 5.4.4:

Write $[\tilde{Y}, \tilde{X}]^v = L_\beta J$. So $\|[\tilde{Y}, \tilde{X}]^v\|^2 = \|L_\beta J\|^2 = \langle\langle L_\beta J, L_\beta J\rangle\rangle$. Let $\alpha_J(\beta) = L_\beta J$. α_J is now a map from C^∞ vector fields $\mathcal{X}(M)$ on M to C^∞ 1-1 tensors $C^\infty(T_1^1 M)$ on M. As such a map, α_J has an adjoint α_J^* (cf. discussion in section 2.5), namely for symmetric trace free 1-1 tensors A, in conformal coordinates $g_{ij} = p\delta_{ij}$,

$$\alpha_J^* : C^\infty\left(T_1^1 M\right) \to \mathcal{X}(M)$$

$$\alpha_J^*(A) = \left(-\frac{2}{p}(\delta_g A)_2, \; +\frac{2}{p}(\delta_g A)_1\right) \quad.$$

Thus if $\beta = (\beta^1, \beta^2)$ in conformal coordinates

$$\|L_\beta J\|^2 = \|\alpha_J \beta\|^2 = \langle\langle \alpha_J\beta, \alpha_J\beta\rangle\rangle = \langle \alpha_J^* \alpha_J \beta, \beta\rangle_{L^2} \quad,$$

where $\langle \cdot, \cdot \rangle_{L^2}$ as in section 1.4 denotes the g inner product of vector fields on M.

But from theorem 5.3.1 it follows that

$$\alpha_J^*(\alpha_J\beta) = \left(-\frac{2}{p}\frac{\partial}{\partial y}(ad - bc), \; -\frac{2}{p}\frac{\partial}{\partial x}(-ad + bc)\right) \quad.$$

Therefore

$$\langle \alpha_J^* \alpha_J \beta, \beta\rangle_{L^2} = 2\int_M \left[-p\beta^1\frac{\partial}{\partial y}(ad - bc) + p\beta^2\frac{\partial}{\partial x}(ad - bc)\right] dx\, dy$$

integrating by parts we see that this is equal to

$$2\int_M (ad - bc)\frac{1}{p}\left\{\frac{\partial}{\partial y}(p\beta^1) - \frac{\partial}{\partial x}(p\beta^2)\right\} d\mu_{g(J)} \qquad (5.30)$$

$$= 2\int_M (ad - bc)\left\{\delta_g(J\beta)\right\} d\mu_{g(J)} \quad.$$

Now $L_{J\beta}J = JL_\beta J = J[\tilde{Y}, \tilde{X}]^v$ and

$$\delta_g\left(J[\tilde{Y}, \tilde{X}]^v\right) = *\delta_g[\tilde{Y}, \tilde{X}]^v = *d\lambda \quad.$$

Therefore by lemma 5.4.5

$$\delta_g(J\beta) \;=\; \rho$$
$$\Delta\rho - \rho \;=\; \Delta\lambda \quad.$$

Thus (5.30) is equal to

$$2\int_M \lambda\rho\, d\mu_g \quad.$$

The operator $\mathcal{L}$ is clearly strictly negative and selfadjoint. So $\mathcal{L}\rho = (\mathcal{L} + I)\lambda$ and hence

$$\rho = \mathcal{L}^{-1}(\mathcal{L} + I)\lambda = \lambda + \mathcal{L}^{-1}\lambda$$

and

$$\int_M \rho\lambda d\mu_g = \int_M \lambda^2 d\mu_g + \int_M (\mathcal{L}^{-1}\lambda)\lambda d\mu_g \quad.$$

This concludes the proof of theorem 5.4.4.   ∎

We are now ready to complete the computation of the induced curvature of Teichmüller space.

Lemma 5.4.6

$$\left\langle\!\left\langle \nabla_{\tilde{Y}}\tilde{X}, (\nabla_{\tilde{Y}}\tilde{X} - \nabla_{\tilde{X}}\tilde{Y})^v \right\rangle\!\right\rangle = -\left\langle\!\left\langle \nabla_{\tilde{X}}\tilde{Y}, (\nabla_{\tilde{Y}}\tilde{X} - \nabla_{\tilde{X}}\tilde{Y})^v \right\rangle\!\right\rangle = \frac{1}{2}\left\|[\tilde{X}, \tilde{Y}]^v\right\|^2 .$$

PROOF: Apply lemma 5.4.3 to both the left side and the first term of the right side of the equality. ∎

As an immediate consequence we have

Lemma 5.4.7

$$\left\|(\nabla_{\tilde{Y}}\tilde{X})^v\right\|^2 = \left\|(\nabla_{\tilde{X}}\tilde{Y})^v\right\|^2 .$$

∎

Consider now the expression (5.29) for the curvature. We can now evaluate the second term $\left\langle\!\left\langle (\nabla_{\tilde{Y}}\tilde{Y})^v, (\nabla_{\tilde{X}}\tilde{X})^v \right\rangle\!\right\rangle$.

Lemma 5.4.8 *Let $T = L_\beta J$ be a vertical field, and let $\tilde{Y}$ be the horizontal lift of a vector field Y on $\mathcal{T}(M)$. Then*

$$\left\langle\!\left\langle \nabla_{\tilde{Y}}\tilde{Y}, T \right\rangle\!\right\rangle_J = +\frac{1}{2}\int\limits_M tr(\tilde{Y}^2)\delta_g\beta d\mu_{g(J)} .$$

PROOF: This follows immediately from an application of theorem 5.3.5. As a consequence we have

Lemma 5.4.9

$$\left\langle\!\left\langle (\nabla_{\tilde{Y}}\tilde{Y})^v, (\nabla_{\tilde{X}}\tilde{X})^v \right\rangle\!\right\rangle = \frac{1}{2}\int\limits_M (c^2 + d^2)\rho d\mu_g = \frac{1}{2}\int\limits_M (a^2 + b^2)\theta d\mu_g$$

where again in conformal coordinates we represent $\tilde{Y}$ and $\tilde{X}$ be the matrices $\begin{pmatrix} c & d \\ d & -c \end{pmatrix}$ and $\begin{pmatrix} a & b \\ b & -a \end{pmatrix}$ respectively, where

$$\Delta\rho - \rho = a^2 + b^2, \quad \Delta\theta - \theta = c^2 + d^2 .$$

Thus

$$\mathcal{L}\rho = (\mathcal{L} + I)(a^2 + b^2)$$
$$\mathcal{L}\theta = (\mathcal{L} + I)(c^2 + d^2)$$

or

$$\rho = (a^2 + b^2) + \mathcal{L}^{-1}(a^2 + b^2)$$

and

$$\theta = (c^2 + d^2) + \mathcal{L}^{-1}(c^2 + d^2) \ .$$

As a consequence we can obtain an explicit expression for the second term of formula (5.29), namely:

Theorem 5.4.10

$$\left\langle\left\langle (\nabla_{\tilde{Y}}\tilde{Y})^v, (\nabla_{\tilde{X}}\tilde{X})^v \right\rangle\right\rangle = \frac{1}{2}\int_M (a^2 + b^2)(c^2 + d^2)d\mu_g + \frac{1}{2}\int_M \left(\mathcal{L}^{-1}(a^2 + b^2)\right)(c^2 + d^2)d\mu_g \ .$$

Theorem 5.4.11

$$\frac{1}{2} \cdot \left\|(\nabla_{\tilde{Y}}\tilde{X} + \nabla_{\tilde{X}}\tilde{Y})^v\right\|^2 = \int_M (ac + bd)^2 d\mu_{g(J)} + \int_M \{\mathcal{L}^{-1}(ac + bd)\}(ac + bd)d\mu_{g(J)} \ .$$

PROOF: Let $H = \nabla_{\tilde{Y}}\tilde{X} + \nabla_{\tilde{X}}\tilde{Y}$.
Then $H^v = L_\beta J$, where if $\rho = -\delta_g\beta$, then using formula (5.16) it follows as in lemma 5.4.5 that

$$\Delta\rho - \rho = \Delta\mu$$

where $\mu = \frac{1}{4}trace(\tilde{X}\tilde{Y} + \tilde{Y}\tilde{X})$. The proof now proceeds exactly as in theorem 5.4.4. ∎

Lemma 5.4.12

$$\left\|(\nabla_{\tilde{Y}}\tilde{X})^v\right\|^2 = \frac{1}{4}\left\{\left\|(\nabla_{\tilde{Y}}\tilde{X} + \nabla_{\tilde{X}}\tilde{Y})^v\right\|^2 + \left\|(\nabla_{\tilde{Y}}\tilde{X} - \nabla_{\tilde{X}}\tilde{Y})^v\right\|^2\right\} \ .$$

PROOF: $2\nabla_{\tilde{Y}}\tilde{X} = (\nabla_{\tilde{Y}}\tilde{X} + \nabla_{\tilde{X}}\tilde{Y}) + (\nabla_{\tilde{Y}}\tilde{X} - \nabla_{\tilde{X}}\tilde{Y})$. Take the normal component and then the square of the norm of both sides and apply lemma 5.4.7 to get the result. ∎

Using this formula for the third term of formula (5.29) for the curvature we can calculate it explicitly using theorem 5.4.11. Thus we have:

Theorem 5.4.13

$$-\|(\nabla_{\tilde{Y}} \tilde{X})^v\|^2 \;=\; -\tfrac{1}{2} \int_M (ac+bd)^2 d\mu_g - \tfrac{1}{2} \int_M (ad-bc)^2 d\mu_g$$
$$-\tfrac{1}{2} \int_M \{\mathcal{L}^{-1}(ac+bd)\}(ac+bd)d\mu_g$$
$$-\tfrac{1}{2} \int_M \{\mathcal{L}^{-1}(ad-bc)\}(ad-bc)d\mu_g \;.$$

Having computed every term in formula (5.29) we can add them to obtain our final formula for the sectional curvature $\mathcal{K}(X,Y)$ of Teichmüller space $\mathcal{T}(M)$, with respect to the Weil-Petersson metric namely adding the results of formulas (5.12), (5.13) and theorems 5.4.4, 5.4.10 and 5.4.13 we find that

$$
\begin{aligned}
2 \cdot \mathcal{K}(X,Y) \;=\; & -2 \int_M (ad-bc)^2 d\mu_g \\
& +\tfrac{1}{2} \int_M (a^2+b^2)(c^2+d^2)d\mu_g \\
& +\tfrac{1}{2} \int_M \{\mathcal{L}^{-1}(a^2+b^2)\}(c^2+d^2)d\mu_g \\
& -\tfrac{1}{2} \int_M (ac+bd)^2 d\mu_g - \tfrac{1}{2} \int_M (ad-bc)^2 d\mu_g \\
& -\tfrac{1}{2} \int_M \mathcal{L}^{-1}(ac+bd)(ac+bd)d\mu_g \\
& -\tfrac{1}{2} \int_M \{\mathcal{L}^{-1}(ad-bc)\}(ad-bc)d\mu_g \\
& +2 \int_M (ad-bc)^2 d\mu_g + 2 \int_M \{\mathcal{L}^{-1}(ad-bc)\}(ad-bc)d\mu_g \\
\;=\; & \tfrac{1}{2} \int_M F \, d\mu_g \\
& +\tfrac{1}{2} \int_M \{\mathcal{L}^{-1}(a^2+b^2)\}(c^2+d^2)d\mu_g \\
& -\tfrac{1}{2} \int_M \{\mathcal{L}^{-1}(ad-bc)\}(ad-bc)d\mu_g \\
& -\tfrac{1}{2} \int_M \{\mathcal{L}^{-1}(ac+bd)\}(ac+bd)d\mu_g \\
& +2 \int_M \{\mathcal{L}^{-1}(ad-bc)\}(ad-bc)d\mu_g
\end{aligned}
$$

where

$$F = \left\{(a^2+b^2)(c^2+d^2) - (ac+bd)^2 - (ad-bc)^2\right\} \equiv 0 \;.$$

We therefore obtain our main results:

Theorem 5.4.14 *Let X and Y be WP-orthonormal vector fields on Teichmüller's space $\mathcal{A}/\mathcal{D}_0$ and $\tilde{X}, \tilde{Y}$ their horizontal lifts to the bundle $\mathcal{A}$. Then if*

$$\lambda = \frac{1}{4} trace\left(J(\tilde{X}\tilde{Y} - \tilde{Y}\tilde{X})\right) \;, \quad \mu = \frac{1}{4} trace(\tilde{X}\tilde{Y} + \tilde{Y}\tilde{X}) \;,$$

the sectional curvature $\mathcal{K}$ of $\mathcal{A}/\mathcal{D}_0$ with respect to its Weil-Petersson metric is given by

$$\mathcal{K}(X,Y) \;=\; +\frac{3}{4}\int_M (\mathcal{L}^{-1}\lambda)\lambda\, d\mu_g - \frac{1}{4}\int_M (\mathcal{L}^{-1}\mu)\mu\, d\mu_g \qquad (5.31)$$
$$+\frac{1}{4}\int_M \left(\mathcal{L}^{-1}(a^2+b^2)\right)(c^2+d^2)d\mu_g \;.$$

Theorem 5.4.15 *The holomorphic sectional curvature of Teichmüller's space is strictly negative and bounded above by* $-1/8\pi(p-1)$, *where* $p = genus(M)$.

PROOF: Let $Y = JX$. Then $\lambda = a^2 + b^2 = c^2 + d^2$, and

$$\mathcal{K}(X,Y) = +\int_M (\mathcal{L}^{-1}\lambda)\lambda\, d\mu_g < 0$$

since the elliptic operator $\mathcal{L}$ is strictly negative. The fact that the Weil-Petersson norm of X is 1 implies

$$1 = \|X\|_{WP}^2 = \int_M \lambda d\mu_g \;.$$

Now let $\lambda = \mathcal{L}f = \Delta f - f$. So $\int_M \lambda = -\int_M f = 1$. Therefore

$$\mathcal{K}(X,Y) = \int_M (\mathcal{L}f)f\, d\mu_g \leq -\int_M f^2\, d\mu_g \;.$$

But

$$1 \leq \int_M 1\, d\mu_g \cdot \int_M f^2\, d\mu_g \;,$$

which implies that $-\int_M f^2 \leq -1/\mathrm{vol}(M)$. By the Gauss-Bonnet theorem 1.3.8

$$\mathrm{vol}(M) = \int_M d\mu_g = 4\pi(2p-2)$$

and the result follows immediately.

Remark 5.4.1 *This bound differs from results other authors have found [125],[99],[57] in that we take* $\mathcal{T}(M) = \mathcal{M}_{-1}/\mathcal{D}_0$ *where* $\mathcal{M}_{-1}$ *are those metrics of scalar curvature -1, whereas other authors take* $\mathcal{T}(M) = \mathcal{M}_{-2}/\mathcal{D}_0$, *i.e. metrics whose Gauss curvature is -1.*

Finally, to see that the sectional curvature is negative, we need the following lemma. Using the uniformization theorem we can represent M with a given conformal structure as $\mathbb{H}/\Gamma$, $\mathbb{H}$ the hyperbolic upper half plane, Γ a subgroup of $SL_2(\mathbb{R})$. Then from the fact that the Green's function of $-\mathcal{L}$ on a fundamental domain is positive and Hölder's inequality we obtain

Lemma 5.4.16 *For any C^2-functions ξ, η on M we have*

$$\left|\mathcal{L}^{-1}(\xi\eta)\right| \leq \left|-\mathcal{L}^{-1}\xi^2\right|^{1/2}\left|-\mathcal{L}^{-1}\eta^2\right|^{1/2}.$$

Applying this lemma and Hölder's inequality term by term to the formula (5.31) we see that

$$-\int_M (\mathcal{L}^{-1}\mu)\mu \leq -\int_M \left\{\mathcal{L}^{-1}(a^2 + b^2)\right\}(c^2 + d^2)d\mu_g.$$

This immediately implies our final results

Theorem 5.4.17 *The sectional curvature of Teichmüller space with respect to its Weil-Petersson metric is negative,*

and with a bit of further work

Theorem 5.4.18 *The Ricci curvature of Teichmüller space is strictly negative and bounded above by*

$$\frac{-1}{8\pi(p-1)}.$$

5.5 An Asymptotic Property of Weil-Petersson Geodesics

We have already mentioned that the Weil-Petersson metric is not complete [126]. Therefore by the theorem of Hopf and Rinow [76] there must exist geodesics which are not defined for all time. In this rather short section we prove the following theorem which shows that if a geodesic is defined for only finite time it must "crash into" a surface of lower genus.

Theorem 5.5.1 *Let $\sigma_{[g]}(t)$, $[g] \in \mathcal{T}(M)$, $t^-[g] < t < t^+[g]$ be a Weil-Petersson geodesic on $\mathcal{T}(M)$ with initial point $[g] = \sigma_{[g]}(0)$, and where $]t^-[g], t^+[g][$ is its maximal domain* *of definition. If $t^+[g] < \infty$ (similarly if $t^-[g] > -\infty$), then for any sequence $t_n \to t^+[g]$ $(t_n \to t^-[g])$ a non-trivial closed geodesic on the surface $\left(M, \sigma_{[g]}(t_n)\right)$ is shrinking in length to zero. In this case we say that $\left(M, \sigma_{[g]}(t_n)\right)$ is developing a node.*

PROOF: Assume the contrary. Then all non-trivial closed geodesics on $\left(M, \sigma_{[g]}(t_n)\right)$ are bounded below in length. Let $\tilde{\sigma}_{[g]}(t_n)$ be the horizontal lift of $\sigma_{[g]}(t)$ to the manifold $\mathcal{M}_{-1}$. $(\pi(\tilde{\sigma}) = \sigma$ and $D\pi\tilde{\sigma}' = \sigma'$, $\pi : \mathcal{M}_{-1} \to \mathcal{T}(M)$.)

By the Mumford Compactness theorem there exists a sequence of diffeomorphisms $f_n \in \mathcal{D}$ such that $f_n^* \tilde{\sigma}_g(t_n)$ converges to $\hat{g} \in \mathcal{M}_{-1}$. Now $\mathcal{D}$ acts as a group of isometries and so the $\mathcal{D}$-action preserves geodesics. Therefore the pull back metrics $f_n^* \tilde{\sigma}_{[g]}(t)$, $t^-[g] < t < t^+[g]$ are also the horizontal lifts $\tilde{\eta}_{[f_n^*g]}(t)$ of a geodesic $\eta_{[f_n^*g]}(t)$ with initial point $[f_n^*g]$, and maximal interval of definition $(t^-[g], t^+[g])$.

By the fundamental existence theorem for ordinary differential equations there is a neighbourhood U of $[\hat{g}] \in \mathcal{T}(M)$ and a $\delta > 0$ such that any geodesic with initial point $[g'] \in U$ is defined for $-\delta < t < \delta$. Now let n be sufficiently large so that $f_n^* \tilde{\sigma}_{[g]}(t_n) \in U$ and $0 < t^+[g] - t_n < \delta/2$. If $g' = \tilde{\eta}_{[f_n^*g]}(t)$ there is a geodesic $\eta_{[g']}(t)$ with

$$\frac{d}{dt} \tilde{\eta}_{[g']}(t)\Big|_{t=0} = \frac{d}{dt} \tilde{\eta}_{[f_n^*g]}(t)\Big|_{t=t_n} .$$

Since $\tilde{\eta}_{[g']}(t) = \tilde{\eta}_{f_n^*g}(t_n + t)$ we have $\tilde{\eta}_{[f_n^*g]}$ defined for $t > t^+[g]$, a contradiction. ∎

We would like to point out here a result of Scott Wolpert [124] relevant to theorem 5.5.1. The theorem of Hopf and Rinow guarantees that if a Riemannian manifold is complete and connected, then any two points in Teichmüller space can be connected by a (minimal) geodesic. The result of Wolpert is, that although the Weil-Petersson metric is not complete, it is nevertheless true that any two points in Teichmüller space can be connected by a Weil-Petersson geodesic (see remark 6.2.6).

6 The Pluri-Subharmonicity of Dirichlet's Energy on $\mathcal{T}(M)$; $\mathcal{T}(M)$ is a Stein Manifold

6.1 Pluri-Subharmonic Functions and Complex Manifolds

Let $\mathcal{F} : N \to \mathbb{R}$ be a real valued function on the complex manifold N. The *Levi-form* of $\mathcal{F}$ at a point $z \in N$ is the complex 2-form

$$\partial\bar{\partial}\mathcal{F} = \frac{\partial^2 \mathcal{F}(z)}{\partial z^\alpha \partial \bar{z}^\beta} dz^\alpha d\bar{z}^\beta \ .$$

If $\xi = \{\xi^\alpha\}$ and $\{\eta^\alpha\} = \eta$ are tangent vectors to N at z then the value of this 2-form on tangent vectors is

$$\frac{\partial^2 \mathcal{F}(z)}{\partial z^\alpha \partial \bar{z}^\beta} \xi^\alpha \bar{\eta}^\beta \ .$$

The usefulness of this form to the study of several complex variables is due to the fundamental work of E.E. Levi.

On a complex manifold the transition maps are holomorphic, and as a consequence of this, if $\frac{\partial^2 \mathcal{F}(z)}{\partial z^\alpha \partial \bar{z}^\beta} \xi^\alpha \bar{\xi}^\beta \geq 0$ in one coordinate system (resp. > 0) then the same holds in all coordinate systems. In this case we say that the sign of $\partial\bar{\partial}\mathcal{F}(z) \geq 0$ (resp > 0) and this is well-defined. A C^2 function $\mathcal{F}$ is *pluri-subharmonic* at z if $sgn\partial\bar{\partial}\mathcal{F}(z) \geq 0$ and strictly pluri-subharmonic at z if $sgn\partial\bar{\partial}\mathcal{F}(z) > 0$. It is called *(strictly) pluri-subharmonic* on N if it is (strictly) pluri-subharmonic at each point $z \in N$.

This definition can be weakened [46] so that one can speak of continuous pluri-subharmonic functions.

The notion of pluri-subharmonic functions is the natural generalization of the notion of subharmonic functions of a single complex variable.

However, the theory of several complex variables differs considerably from that of one complex variable in that many new (and surprising) phenomena arise. For example, although there is a natural analogue in several variables to Riemann's theorem on the removability of isolated singularities for bounded holomorphic functions, much more is true: namely that sufficiently small subsets of $\mathbb{C}^n(n > 1)$, such as isolated points, are automatically removable singularities of holomorphic functions without the requirement that they be bounded.

Another of the surprising characteristics of the theory of several variables is the existence of pairs of open sets $\Omega \subset\subset \Omega' \subseteq \mathbb{C}^n$ such that every function holomorphic in Ω necessarily extends to a function which is holomorphic in the strictly larger set Ω'. *Domains of holomorphy* in $\mathbb{C}^n$ are precisely those domains Ω for which no extension is possible, formally:

Definition 6.1.1 A *domain of holomorphy* in $\mathbb{C}^n$ is an open set Ω for which there exists at least one holomorphic function $f : \Omega \to \mathbb{C}$ which cannot be extended as a holomorphic function through any boundary point of Ω.

It is a fundamental question to find characterizations of such domains. The role of pluri-subharmonic functions in the investigation of such questions was pioneered by the Japanese mathematician K. Oka [88].

For example, the notion of pluri-subharmonicity permits one to define the notion of pseudo convex subset $\Omega \subset \mathbb{C}^n$, as an open set which admits a proper pluri-subharmonic exhaustion function; i.e. a pluri-subharmonic function $f : \Omega \to \mathbb{R}$ such that $\{x \mid f(x) \leq a\}$ is compact. One of the deeper theorems in the theory of several complex variables is the equivalence of the notions of domain of holomorphy and pseudo convex domain [46].

Actually the definition of pseudo convex domain above is rather modern. Another definition of pseudo convexity for domains Ω with smooth boundaries was introduced in 1910 by E.E. Levi [68], namely that $\partial\Omega$ is locally defined as the level set of a pluri-subharmonic function. Ω is then said to be pseudo convex in the sense of Levi. Another basic result in the theory of several complex variables is that an open subset $\Omega \subset \mathbb{C}^n$ with C^2 boundary $\partial\Omega$ is pseudo convex (and hence a domain of holomorphy) iff it is pseudo convex in the sense of Levi.

The set of complex manifolds (or submanifolds of $\mathbb{C}^n$) is quite large. One can ask for restrictions on this set to those manifolds (or complex varieties) that in some sense are generalizations to higher dimensions of open Riemann surfaces. For example, can one have a notion of complex manifold N which guarantees that N admits a sufficiently rich set of holomorphic functions, or permits a generalization of the Mittag-Leffler theorem on the existence of meromorphic functions with prescribed principal parts, or even with prescribed "poles and zeros"? Suitably formulated, these two problems are known in several variables as Cousin's first and second problem, respectively.

The natural complex manifolds (or varieties) which have all these properties are known as *Stein manifolds* for which many (equivalent) definitions are known. We shall take a definition suitable for our work in Teichmüller theory, namely

Definition 6.1.2 A complex manifold N is *Stein* if it admits a pluri-subharmonic exhaustion function $\mathcal{F} : N \rightarrow \mathbb{R}$.

There are several characterizations of Stein manifolds that are worth mentioning, namely

(A) N is Stein, iff

    (i) the holomorphic functions on N separate points

    (ii) for every sequence $\{x_n\} \subset N$ with $x_n \rightarrow \infty$, there exists a holomorphic function f on N with $\lim_{n\rightarrow\infty} |f(x_n)| = \infty$

(B) N is Stein, iff it can be biholomorphically realized as a closed complex submanifold of some $\mathbb{C}^N$ (actually, $N = 2n + 1$, $n = \dim_{\mathbb{C}} N$ suffices).

This is a very rich and important class of manifolds. For example, open domains $\Omega \subset \mathbb{C}^n$ are Stein iff they are domains of holomorphy. For further reading we recommend the books of Gunning [46] and Grauert and Remmert [44].

In section 3 of this chapter we show that Dirichlet's energy $\tilde{E} : T(M) \rightarrow \mathbb{R}^+$ is a C^∞ pluri-subharmonic exhaustion function. As a consequence we have a result first proved by Bers and Ehrenpreis [14] by very different methods:

Theorem 6.1.1 *Teichmüller space* $T(M)$ *is a complex Stein manifold.*

In addition, Lipman Bers [12],[13] has shown that $T(M)$ can be realized as a bounded domain in $\mathbb{C}^N$.

6.2   Dirichlet's Energy is Strictly Pluri-Subharmonic

In order to compute the Levi-form of Dirichlet's energy on Teichmüller space $\mathcal{T}(M)$ one needs a holomorphic coordinate system, say Abresch-Fischer coordinates.

If we view $\tilde{E} : \mathcal{T}(M) \to \mathbb{R}^+$ as a map $\tilde{E} : \mathcal{A}/\mathcal{D}_0 \to \mathbb{R}^+$ we can fix a $[J_0] \in \mathcal{T}(M)$ and let $H = \xi^\alpha H_\alpha$, H_α a basis of $T_{[J_0]}\mathcal{T}(M)$ invariant under the almost complex structure $\hat{\Phi}$ of $\mathcal{T}(M)$ and a basis which also determines the local coordinate $\{z^\alpha\}$.

If U is a neighborhood of 0 in $T_{[J_0]}\mathcal{T}(M)$ and $\psi : U \to \mathcal{T}(M)$ a holomorphic coordinate system then in these coordinates

$$\frac{\partial^2 \tilde{E}}{\partial z^\alpha \partial \bar{z}^\beta}[J_0]\xi^\alpha \bar{\xi}^\beta = D^2(\tilde{E} \circ \psi)[J_0](H, H) + D^2(\tilde{E} \circ \psi)[J_0](J_0 H, J_0 H) .$$

Such a direct calculation is carried out in appendix E. This calculation, as might be expected, is somewhat involved. However, we are not as interested in an explicit expression of the Levi-form as we are in determining whether or not it is positive or strictly positive. The following result, using Weil-Petersson geodesics allows us to avoid such a detailed calculation. Although this result appears to be generally well known, the proof known to the author is due to Deligne, Griffiths, Morgan and Sullivan [16].

Theorem 6.2.1
For a Hermitian Kähler manifold N with almost complex structure J there always exists a complex coordinate system about each point $p_0 \in N$ such that the Kähler metric G_{ij} in this coordinate system is of the form

$$G_{ij}(p) = \delta_{ij} + O\left(||p - p_0||^2\right) .$$

In particular, all first derivatives of the G_{ij} vanish at p_0.

Remark 6.2.1 *If (N, J) is a complex Hermitian manifold, it is not too difficult to see, that the above condition on the G_{ij} is equivalent to (N, J) being a Hermitian Kähler manifold.*

Remark 6.2.2 *A simple calculation shows that the Abresch-Fischer coordinate system on Teichmüller space $\mathcal{T}(M) = \mathcal{A}/\mathcal{D}_0$ "almost" satisfies the conditions of theorem 6.2.1, namely in these coordinates $G_{ij}(H) = 4\delta_{ij} + O(||H||^2)$.*

Corollary 6.2.2 *In such a coordinate system, as given in the above theorem, all geodesics* $\sigma(t)$ *with* $\sigma(0) = p_0$ *have* $\sigma''(0) = 0$.

PROOF: $\sigma^{k\,\prime\prime}(t) = \Gamma_{ij}^k\left(\sigma(t)\right)\sigma^{i\,\prime}(t)\,\sigma^{j\,\prime}(t)$.

At $t = 0$ we see that

$$\sigma^{k\,\prime\prime}(0) = \Gamma_{ij}^k(p_0)\,\sigma^{i\,\prime}(0)\,\sigma^{j\,\prime}(0)\ .$$

But the Christoffel symbols $\Gamma_{ij}^k(p_0)$ are essentially a linear combination of the derivatives of the G_{ij} at p_0 which necessarily vanish forcing $\sigma^{k\,\prime\prime}(0) = 0$. ∎

Corollary 6.2.3 *In order to determine whether or not the Levi-form of a function* $f : N \to \mathbb{R}$ *at* p_0 *is positive at* p_0, *it suffices to take two geodesics,* $\sigma_1(t)$, $\sigma_2(t)$, *with* $\sigma_1'(0) = H$ *and* $\sigma_2'(0) = J_{p_0}H$, $\sigma_j(0) = p_0$, $j = 1, 2$. *Then if* $H = \xi^\alpha H_\alpha$ *as above, then in the holomorphic coordinate system given by theorem 6.2.1*

$$\frac{\partial^2 f}{\partial z^\alpha \partial \bar{z}^\beta}(p_0)\xi^\alpha \bar{\xi}^\beta = \sum_j \frac{d^2}{dt^2} f\left(\sigma_j(t)\right)\Big|_{t=0}\ .$$

PROOF: A simple calculation. ∎

This last fact will simplify our lives considerably. We know that the map $\tilde{\Psi} : \mathcal{A} \to \mathcal{M}_{-1}$ induces an isometry $\tilde{\Psi}$ between $\mathcal{A}/\mathcal{D}_0$ and $\mathcal{M}_{-1}/\mathcal{D}_0$ with respect to the L^2-metrics and Weil-Petersson metrics on $\mathcal{A}/\mathcal{D}_0$ and $\mathcal{M}_{-1}/\mathcal{D}_0$. This isometry induces a complex structure on $\mathcal{M}_{-1}/\mathcal{D}_0$ which makes $\tilde{\Psi}$ a holomorphic equivalence. As a consequence, in order to compute the sign of the Levi-form it suffices to consider two Weil-Petersson geodesics $\sigma_1(t)$, $\sigma_2(t)$ with $\sigma_j(0) = [g] \in \mathcal{M}_{-1}/\mathcal{D}_0$, $j = 1, 2$ and with $\sigma_1'(0) = h$ and $\sigma_2'(0) = \hat{\Phi}_{[g]}h$, $\hat{\Phi}$ the induced almost complex structure on $T(M)$. Consider now the horizontal lifts $\tilde{\sigma}_j(t)$ of $\sigma_j(t)$ to $\mathcal{M}_{-1}$; i.e. for each t, $\tilde{\sigma}_j'(t)$ is a trace free divergence free 0-2 tensor with respect to the metric $\tilde{\sigma}_j(t) \in \mathcal{M}_{-1}$ and where $\Pi\tilde{\sigma}_j(t) = \sigma_j(t)$, $D\Pi\left(\tilde{\sigma}_j(t)\right)\tilde{\sigma}_j'(t)$, $\Pi : \mathcal{M}_{-1} \to T(M)$.

What does it mean for $\tilde{\sigma}_j(t)$ to be the horizontal lift of a geodesic in $T(M)$? Let $P_g : S_2 \to S_2^{TT}(g)$ be the L^2-orthogonal projection of all 0-2 tensors onto those which are trace free and divergence free with respect to g. Then the geodesic equation of $\sigma(t)$ can be expressed as

$$P_{\tilde{\sigma}_j(t)}\tilde{\sigma}_j''(t) \equiv 0\ . \tag{6.1}$$

Here $\tilde{\sigma}_1'(0) = \tilde{h}$ and $\tilde{\sigma}_2'(0) = \Phi_{[g]}\tilde{h}$ where, by definition, the induced almost complex structure $\Phi_{[g]}$ is given by

$$\Phi_{[g]}h = (Jh^\sharp)_\flat\ ,\quad J = \tilde{\Psi}(g)\ . \tag{6.2}$$

Let us denote the quantity $(Jh^\flat)_\flat$ simply by ih.

Then in some holomorphic coordinate system (given by theorem 6.2.1):

$$\frac{\partial^2 \tilde{E}}{\partial z^\alpha \partial \bar{z}^\beta}[g]\xi^\alpha \bar{\xi}^\beta = \sum_{j=1}^{2} \frac{d^2}{dt^2} \tilde{E}\left(\sigma_j(t)\right)\Big|_{t=0} = \sum_{j=1}^{2} \frac{d^2}{dt^2} \hat{E}\left(\tilde{\sigma}_j(t)\right)\Big|_{t=0} \tag{6.3}$$

where, we recall, the map $\hat{E} : \mathcal{M} \to \mathbb{R}$ is given by

$$\hat{E}(g) = E\left(g, S(g)\right)$$

where $E(g,\cdot)$ is the classical Dirichlet energy and $S(g) : (M,g) \to (M,g_0)$ is the unique harmonic map homotopic to the identity (cf. section 3.1).

The advantage there is that we may now apply the chain rule to the last term in (6.3) to obtain the equation

$$\sum_{j=1}^{2} \frac{d^2}{dt^2} \hat{E}\left(\tilde{\sigma}_j(t)\right)\Big|_{t=0} = \sum_{j=1}^{2} D\hat{E}(g)\sigma_j''(0) + D^2\hat{E}(g)(h,h) + D^2\hat{E}(g)(ih,ih) \ . \tag{6.4}$$

The following lemma illustrates the advantage of using geodesics in our calculation.

Lemma 6.2.4

$$D\hat{E}(g)\tilde{\sigma}_j''(0) \equiv 0, \qquad j = 1,2 \ .$$

PROOF: By lemma 3.1.4 we have the formula

$$D\hat{E}(g)\tilde{\sigma}_j''(0) = -\frac{1}{4}\left\langle\!\left\langle \text{Re}\,\xi(z)dz^2, \left\{\tilde{\sigma}_j''(0)\right\}^T \right\rangle\!\right\rangle \ ,$$

$\text{Re}\,\xi(z)dz^2 \in S_2^{TT}(g)$.

Thus

$$\begin{aligned}
D\hat{E}(g)\tilde{\sigma}_j''(0) &= -\tfrac{1}{4}\langle\!\langle \text{Re}\,\xi(z)dz^2, \tilde{\sigma}_j''(0)\rangle\!\rangle \\
&= -\tfrac{1}{4}\langle\!\langle \text{Re}\,\xi(z)dz^2, P_g\tilde{\sigma}_j''(0)\rangle\!\rangle \\
&\quad -\tfrac{1}{4}\langle\!\langle \text{Re}\,\xi(z)dz^2, (I - P_g)\tilde{\sigma}_j''(0)\rangle\!\rangle \ .
\end{aligned}$$

However $P_g(\tilde{\sigma}_j''(0)) = 0$ since $\sigma_j(t)$ are geodesics. The fact that $I - P_g$ is a projector onto the L^2-orthogonal complement of $S_2^{TT}(g)$ implies that $\langle\!\langle \text{Re}\,\xi(z)dz^2, (I - P_g)\sigma_j''(0)\rangle\!\rangle = 0$ which finishes the proof. ∎

Therefore by (6.4) and our previous remarks the strict positivity of the Levi-forms depends on showing that

$$D^2 \hat{E}(g)(h, h) + D^2 \hat{E}(g)(ih, ih) > 0 \qquad (6.5)$$

for all $h \in T_g \mathcal{M}_{-1}$.

We have already developed formula (3.5) for $D^2 \hat{E}(g)$, namely

$$D^2 \hat{E}(h, h) = \sum_{\alpha=1}^{K} \frac{1}{4} \int_M (h \cdot h) g(x) (\nabla_g S^\alpha, \nabla_g S^\alpha) d\mu_g - \sum_{\alpha=1}^{K} \int_M h^{ij} \frac{\partial W(h)^\alpha}{\partial x^i} \frac{\partial S^\alpha}{\partial x^j} d\mu_g \qquad (6.6)$$

where $W(h) = DS(g)h$, and we assume that we have a fixed isometric embedding of the target manifold (M, g_0) of S into $\mathbb{R}^K$.

The next lemma shows that we can simplify the second term on the right in (6.6).

Lemma 6.2.5

$$-\sum_{\alpha=1}^{K} \int h^{ij} \frac{\partial W(h)^\alpha}{\partial x^i} \cdot \frac{\partial S^\alpha}{\partial x^i} d\mu_g = -D^2 E_g \left(W(h), W(h) \right) \qquad (6.7)$$

where $D^2 E_g$ is the second variation of the Dirichlet energy $S \mapsto E(g, S)$.

PROOF: Since $S(g)$ is a critical point (in fact an absolute minimum) of $S \mapsto E(g, S)$ we know that $\frac{\partial E}{\partial S} \equiv 0$ if evaluated at $(g, S(g))$. Differentiating this identity now with respect to g we obtain

$$\frac{\partial^2 E}{\partial g \partial S} + \frac{\partial^2 E}{\partial S^2} = 0$$

where again these derivatives are evaluated at $(g, S(g))$, thus

$$\begin{aligned}
-D^2 E_g \left(W(h), W(h) \right) &= -\frac{\partial^2 E}{\partial S^2} (W(h), W(h)) = \frac{\partial^2 E}{\partial g \partial S} (h, W(h)) \\
&= -\sum_{\alpha=1}^{K} \int_M h^{ij} \frac{\partial W(h)^\alpha}{\partial x^i} \cdot \frac{\partial S^\alpha}{\partial x^j} d\mu_g
\end{aligned}$$

which concludes 6.2.5. ∎

Therefore using the fact that $(ih \cdot ih) = h \cdot h$ we obtain the following formula:

$$\begin{aligned}
D^2 \hat{E}(h, h) + D^2 \hat{E}(ih, ih) &= \frac{1}{2} \sum_{\alpha=1}^{K} \int (h \cdot h) g(x) (\nabla_g S^\alpha, \nabla_g S^\alpha) d\mu_g \\
&\quad - D^2 E_g (W(h), W(h)) \\
&\quad - D^2 E_g (W(ih), W(ih))
\end{aligned} \qquad (6.8)$$

where the second variation of Dirichlet's energy at $S(g)$ is given by

$$
\begin{aligned}
D^2 E_g(W, W) \;=\; & \int_M \left\{ \left\| \nabla_{\frac{\partial}{\partial x}} W \right\|^2 + \left\| \nabla_{\frac{\partial}{\partial y}} W \right\|^2 \right\} dx\, dy \\
& - \sum_j \int g_0 \left(\mathcal{R}\left(\frac{\partial S}{\partial x^j}, W \right) W, \frac{\partial S}{\partial x^j} \right) d\mu_g
\end{aligned}
\tag{6.9}
$$

where the local coordinates (x^1, x^2) are written as (x, y). g_0 represents the curvature of (M, g_0), and $\nabla_{\frac{\partial}{\partial x^\alpha}}$ represents covariant differentiation "along" S. Our goal is now to show that (6.8) is *strictly* positive.

Remark 6.2.3 *Expression (6.8) is "essentially" (up to a constant multiple) the Levi-form of Dirichlet's energy in Abresch-Fischer coordinates (see remark 6.2.2). In appendix E we give a direct calculation of this fact.*

For simplicity of notation denote $D^2 E_g$ by Q, so $Q(V, W) = D^2 E_g(V, W)$. V and W vector fields along S, i.e. $V(x) \in T_{S(g)(x)} M$.

We now want a formula for $Q(W(h), V)$.

Since we are assuming that (M, g_0) is isometrically embedded in $\mathbb{R}^K$ we may let $\Pi(x)$ be the orthogonal projection of $\mathbb{R}^K$ onto $T_x M$. Then the condition that $S : (M, g) \to (M, g_0)$ is harmonic can be written extrinsically as

$$
\Pi(S(x)) \Delta_g S = 0
\tag{6.10}
$$

where

$$
\Delta_g S = \frac{1}{\sqrt{g}} \frac{\partial S}{\partial x^i} \left(\sqrt{g}\, g^{ij} \frac{\partial S}{\partial x^i} \right)
\tag{6.11}
$$

is the Laplace-Beltrami operator on the vector valued map $S : M \to \mathbb{R}^K$. Differentiating (6.10) in the direction of a trace free (with respect to g) h, we obtain:

$$
D\Pi(s)\left[W(h)\right]\Delta_g S + \Pi(s)\Delta_g W(h) + \Pi(s)\left\{ \frac{\partial}{\partial g} \Delta_g \right\}(h) S = 0 \; .
\tag{6.12}
$$

By formula (6.11) we see that

$$
\left\{ \frac{\partial}{\partial g} \Delta_g \right\}(h) S = -\frac{1}{\sqrt{g}} \frac{\partial}{\partial x^i} \left(\sqrt{g}\, h^{ij} \frac{\partial S}{\partial x^j} \right) \; .
\tag{6.13}
$$

But necessarily the second variation $Q\left(W(h),V\right)$ equals

$$Q\left(W(h),V\right) = -\int_M \langle D\text{II}(s)\left[W(h)\right]\Delta_g S + \text{II}(S)\Delta_g W(h), V\rangle_{euc}\, d\mu_g\ ,$$

where $\langle \cdot,\cdot\rangle_{euc}$ is the $\mathbb{R}^K$ euclidian inner product. For simplicity we drop the subscript *euc*, and this should cause no confusion.

Now writing W for $W(h)$, taking the euclidian inner product of (6.13) with V and integrating by parts we obtain

$$\begin{aligned}
Q\left(W(h),V\right) &= \int_M \left\langle h^{11}\frac{\partial S}{\partial x} + h^{21}\frac{\partial S}{\partial y}, \frac{\partial V}{\partial x}\right\rangle \sqrt{g}\, dx\, dy \\
&+ \int_M \left\langle h^{22}\frac{\partial S}{\partial y} + h^{12}\frac{\partial S}{\partial x}, \frac{\partial V}{\partial y}\right\rangle \sqrt{g}\, dx\, dy\ .
\end{aligned} \tag{6.14}$$

Since

$$\begin{aligned}
\nabla_{\frac{\partial}{\partial x}} V &= \text{II}(S)\tfrac{\partial V}{\partial x} \\
\nabla_{\frac{\partial}{\partial y}} V &= \text{II}(S)\tfrac{\partial V}{\partial y}
\end{aligned}$$

we may rewrite (6.14) as

$$\begin{aligned}
Q\left(W(h),V\right) &= \int_M \left\langle h^{11}\frac{\partial S}{\partial x} + h^{21}\frac{\partial S}{\partial y}, \nabla_{\frac{\partial}{\partial x}} V\right\rangle \sqrt{g}\, dx\, dy \\
&+ \int_M \left\langle h^{22}\frac{\partial S}{\partial y} + h^{12}\frac{\partial S}{\partial x}, \nabla_{\frac{\partial}{\partial y}} V\right\rangle \sqrt{g}\, dx\, dy\ .
\end{aligned} \tag{6.15}$$

Using the complex notation for conformal coordinates on (M,g) and (M,g_0) with $(g_0)_{ij} = \rho\delta_{ij}$ we define $\nabla_{\frac{\partial}{\partial z}} = \frac{1}{2}\left\{\nabla_{\frac{\partial}{\partial x}} - i\nabla_{\frac{\partial}{\partial y}}\right\}$ and $\nabla_{\frac{\partial}{\partial \bar z}} = \frac{1}{2}\left\{\nabla_{\frac{\partial}{\partial x}} + i\nabla_{\frac{\partial}{\partial y}}\right\}$. Then a computation shows that (6.15) can be written

$$Q\left(W(h),V\right) = \text{Re}\, 2\int_M \rho\left\{S_z \overline{\nabla_{\frac{\partial}{\partial \bar z}} V} + \overline{S_z}\nabla_{\frac{\partial}{\partial \bar z}} V\right\} H\, dx\, dy \tag{6.16}$$

where $H = h_1^1 + ih_2^1$, Re denotes real part and V denotes *any* smooth vector field along S. See [114] for a detailed calculation in local coordinates.

Remark 6.2.4 *Formula (6.16) gives us a very quick and elegant proof of (iii) of theorem 3.1.3, namely that $D^2\tilde E[g_0] = \langle\cdot,\cdot\rangle_{WP}$, for $\tilde E : T(M) \to \mathbb{R}$ Dirichlet's energy. The main difficulty in the proof of this result was to show that $DS(g_0)h = W(h) = 0$ (lemma 3.1.5). Since the second variation Q of Dirichlet's energy for harmonic maps is positive definite,* *to show that* $W(h) = 0$ *it suffices to prove that* $Q(W(h), V) = 0$ *for all* V, *and this follows relatively easily from (6.16). To see this, note that when* $g = g_0$, $S = id$ *and therefore* $S_z = 1$ *and* $S_{\bar{z}} = 0$. *Moreover, in conformal coordinates* $g_{ij} = \rho \delta_{ij} = (g_0)_{ij}$,

$$\nabla_{\frac{\partial}{\partial \bar{z}}} V = \frac{\partial V}{\partial \bar{z}} + \frac{\rho s}{\rho} S_z V = \frac{\partial V}{\partial \bar{z}} \qquad \text{and} \qquad \rho H = h \ ,$$

h *a holomorphic quadratic differential. With these considerations, (6.16) becomes*

$$Q(W(h), V) = \operatorname{Re} 2 \int_M \overline{\frac{\partial V}{\partial \bar{z}}} h \, dx \, dy = 2 \int_M \left\langle \frac{\partial V}{\partial \bar{z}}, h \right\rangle \, dx \, dy \ .$$

Integrating by parts and using the fact that $\frac{\partial h}{\partial \bar{z}} = 0$, *we see that* $Q(W(h), V) = 0$ *for every* V.

From (6.8) it follows that Dirichlet's energy is strictly pluri-subharmonic if we can show the inequality

$$Q\left(W(h), W(h)\right) + Q\left(W(ih), W(ih)\right) < \frac{1}{2} \sum_{\alpha=1}^{K} \int (h \cdot h) g(x) (\nabla_g S^\alpha, \nabla_g S^\alpha) d\mu_g \qquad (6.17)$$

where Q *is given by (6.16). This is what we now proceed to do. Our computational trick will be to write the left hand side of (6.17) as follows:*

$$2 \left\{ Q\left(W(h), W(h)\right) + Q\left(W(ih), W(ih)\right) \right\} = \begin{aligned} & Q\left(W(ih), W(ih) + iW(h)\right) \\ & + Q\left(W(h), W(h) - iW(ih)\right) \\ & + Q\left(W(ih), W(ih) - iW(h)\right) \\ & + Q\left(W(h), W(h) + iW(ih)\right) \end{aligned}$$

Now

$$Q\left(W(ih), W(ih) + iW(h)\right) + Q\left(W(h), W(h) - iW(ih)\right)$$

is of the form

$$Q\left(W(ih), iV\right) + Q\left(W(h), V\right) \qquad (6.18)$$

where $V = W(h) - iW(ih)$.

Since

$$Q\left(W(h), V\right) = \operatorname{Re} 2 \int_M \rho \left\{ S_z \overline{\nabla_{\frac{\partial}{\partial \bar{z}}} V} + \overline{S_z} \nabla_{\frac{\partial}{\partial \bar{z}}} V \right\} H \, dx \, dy$$

expression (6.18) can be written as

$$Q\left(W(ih), iV\right) + Q\left(W(h), V\right) = \operatorname{Re} 4 \int_M \rho \cdot S_z \overline{\nabla_{\frac{\partial}{\partial \bar{z}}} V} \cdot H \, dx \, dy \ . \qquad (6.19)$$

Moreover

$$Q\left(W(ih),W(ih)-iW(h)\right)+Q\left(W(h),W(h)+iW(ih)\right)$$

is of the form

$$Q\left(W(ih),-iU\right)+Q\left(W(h),U\right)$$

where $U=W(h)+iW(ih)$ and hence equal to

$$\mathrm{Re}\,4\int_M \rho\,\overline{S_z}\,\nabla_{\frac{\partial}{\partial \bar z}}U\,H\,dx\,dy\ . \tag{6.20}$$

Combining (6.19) and (6.20) we obtain the desired formula:

$$\begin{aligned}
2\left\{Q\left(W(ih),W(ih)\right)+Q\left(W(h),W(h)\right)\right\}=\\
=\mathrm{Re}\,4\int_M \rho\left\{S_z\overline{\nabla_{\frac{\partial}{\partial \bar z}}\{W(h)-iW(ih)\}}\right\}H\,dx\,dy\\
+\mathrm{Re}\,4\int_M \rho\left\{\overline{S_z}\nabla_{\frac{\partial}{\partial \bar z}}\{W(h)+iW(ih)\}\right\}H\,dx\,dy\ .
\end{aligned} \tag{6.21}$$

Applying Schwarz's inequality we see that

$$\begin{aligned}
2\left\{Q\left(W(ih),W(ih)\right)+Q\left(W(h),W(h)\right)\right\}\le\\
\le\quad 2\int_M \rho\left\{|S_z|^2+|S_z|^2\right\}|H|^2 dx\,dy\\
+2\int_M \rho\left|\nabla_{\frac{\partial}{\partial \bar z}}(W(h)-iW(ih))\right|^2 dx\,dy\\
+2\int_M \rho\left|\nabla_{\frac{\partial}{\partial \bar z}}(W(h)+iW(ih))\right|^2 dx\,dy\\
=\quad \frac{1}{2}\sum_{\alpha=1}^{K}\int_M (h\cdot h)g(x)(\nabla_g S^\alpha,\nabla_g S^\alpha)d\mu_g\\
+2\int_M \rho\left\{\left|\nabla_{\frac{\partial}{\partial \bar z}}W(h)\right|^2+\left|\nabla_{\frac{\partial}{\partial z}}W(h)\right|^2\right\}dx\,dy\\
+2\int_M \rho\left\{\left|\nabla_{\frac{\partial}{\partial \bar z}}W(ih)\right|^2+\left|\nabla_{\frac{\partial}{\partial z}}W(ih)\right|^2\right\}dx\,dy\\
+4\int_M \left\langle\nabla_{\frac{\partial}{\partial \bar z}}iW(h),\nabla_{\frac{\partial}{\partial \bar z}}W(ih)\right\rangle dx\,dy\\
-4\int_M \left\langle\nabla_{\frac{\partial}{\partial \bar z}}iW(h),\nabla_{\frac{\partial}{\partial z}}W(ih)\right\rangle dx\,dy\ ,
\end{aligned} \tag{6.22}$$

where $\langle\xi,\eta\rangle:=\rho\cdot\mathrm{Re}\,\xi\bar\eta$ again represents the Riemannian inner product on (M,g_0) which is the inner product on $\mathbb{R}^K$.

Let us concentrate on these last two terms. Integrating by parts we see that their sum equals

$$
-4\left\{\int_M \left\langle i\left\{\nabla_{\frac{\partial}{\partial z}}\nabla_{\frac{\partial}{\partial \bar z}}W(h) - \nabla_{\frac{\partial}{\partial \bar z}}\nabla_{\frac{\partial}{\partial z}}W(h)\right\}, W(ih)\right\rangle\right\}
$$

$$
= 2\int_M \left\langle \left(\nabla_{\frac{\partial}{\partial x}}\nabla_{\frac{\partial}{\partial y}} - \nabla_{\frac{\partial}{\partial y}}\nabla_{\frac{\partial}{\partial x}}\right) W(h), W(ih)\right\rangle dx\, dy \ .
$$

But this last term is the integral of a curvature term, namely

$$
2\int_M \left\langle \mathcal{R}\left(\frac{\partial S}{\partial x}, \frac{\partial S}{\partial y}\right) W(h), W(ih)\right\rangle dx\, dy \ ,
$$

where $\mathcal{R}$ is the curvature tensor of (M, g_0). We therefore arrive at the inequality

$$
2\left\{Q\left(W(ih), W(ih)\right) + Q\left(W(h), W(h)\right)\right\} \le
$$

$$
\le \quad \frac{1}{2}\sum_{\alpha=1}^{K}\int_M (h\cdot h)g(x)(\nabla_g S^\alpha, \nabla_g S^\alpha)d\mu_g
$$

$$
+ 2\int_M \rho\left\{\left|\nabla_{\frac{\partial}{\partial z}}W(h)\right|^2 + \left|\nabla_{\frac{\partial}{\partial \bar z}}W(h)\right|^2\right\} dx\, dy
$$

$$
+ 2\int_M \rho\left\{\left|\nabla_{\frac{\partial}{\partial z}}W(ih)\right|^2 + \left|\nabla_{\frac{\partial}{\partial \bar z}}W(ih)\right|^2\right\} dx\, dy
$$

$$
+ 2\int_M \left\langle \mathcal{R}\left(\frac{\partial S}{\partial x}, \frac{\partial S}{\partial y}\right) W(h), W(ih)\right\rangle dx\, dy \ .
$$

This is further equal to:

$$
\frac{1}{2}\sum_{\alpha=1}^{K}\int_M (h\cdot h)g(x)(\nabla_g S^\alpha, \nabla_g S^\alpha)d\mu_g
$$

$$
+ \int_M \rho\left\{\left|\nabla_{\frac{\partial}{\partial x}}W(h)\right|^2 + \left|\nabla_{\frac{\partial}{\partial y}}W(h)\right|^2\right\} dx\, dy
$$

$$
+ \int_M \rho\left\{\left|\nabla_{\frac{\partial}{\partial x}}W(ih)\right|^2 + \left|\nabla_{\frac{\partial}{\partial y}}W(ih)\right|^2\right\} dx\, dy
$$

$$
+ 2\int_M \left\langle \mathcal{R}\left(\frac{\partial S}{\partial x}, \frac{\partial S}{\partial y}\right) W(h), W(ih)\right\rangle dx\, dy
$$

$$
= 2\left\{Q\left(W(ih), W(ih)\right) + Q\left(W(h), W(h)\right)\right\} \ .
$$

Taking into account formula (6.9) for the second variation Q of Dirichlets energy E_g, we may bring the first two integrals above to the right hand side to obtain a new inequality, namely:

$$Q\left(W(ih), W(ih)\right) + Q\left(W(h), W(h)\right)$$
$$- \int_M \left\langle \mathcal{R}\left(W(ih), \frac{\partial S}{\partial x^l}\right) \frac{\partial S}{\partial x^l}, W(ih)\right\rangle dx\, dy$$
$$- \int_M \left\langle \mathcal{R}\left(W(h), \frac{\partial S}{\partial x^l}\right) \frac{\partial S}{\partial x^l}, W(h)\right\rangle dx\, dy \tag{6.23}$$
$$\leq \quad \frac{1}{2} \sum_{\alpha=1}^{K} \int_M (h \cdot h) g(x)(\nabla_g S^\alpha, \nabla_g S^\alpha) d\mu_g$$
$$+ 2 \int_M \left\langle \mathcal{R}\left(\frac{\partial S}{\partial x}, \frac{\partial S}{\partial y}\right) W(h), W(ih)\right\rangle dx\, dy \ .$$

Since (M, g_0) has negative curvature -1 and S is a diffeomorphism, the sum of the curvature terms on the left hand side of (6.23) is non-negative; it is strictly positive if either $W(h)$ or $W(ih)$ is non-zero.

Our next step is to show that the curvature terms on the left of the inequality (6.23) can be used to "absorb" the curvature term on the right hand side. In doing this we shall again explicitly use the fact that the range of S is two dimensional and that S is a diffeomorphism.

With these facts in mind we may write

$$\begin{aligned} W(ih) &= c\frac{\partial S}{\partial x} + d\frac{\partial S}{\partial y}\\ W(h) &= a\frac{\partial S}{\partial x} + b\frac{\partial S}{\partial y} \end{aligned}$$

for uniquely determined functions a, b, c, d on M. It follows that

$$- \int_M \left\langle \mathcal{R}\left(W(ih), \frac{\partial S}{\partial x^l}\right) \frac{\partial S}{\partial x^l}, W(ih)\right\rangle dx\, dy$$
$$- \int_M \left\langle \mathcal{R}\left(W(h), \frac{\partial S}{\partial x^l}\right) \frac{\partial S}{\partial x^l}, W(h)\right\rangle dx\, dy$$
$$= -\int_M (a^2 + b^2 + c^2 + d^2)\left\langle \mathcal{R}\left(\frac{\partial S}{\partial x}, \frac{\partial S}{\partial y}\right) \frac{\partial S}{\partial y}, \frac{\partial S}{\partial x}\right\rangle dx\, dy$$

and this is > 0 if either $W(ih)$ or $W(h)$ are non-zero. It also follows that

$$2\int_M \left\langle \mathcal{R}\left(\frac{\partial S}{\partial x}, \frac{\partial S}{\partial y}\right) W(h), W(ih)\right\rangle dx\, dy = -2\int_M (ad - bc)\left\langle \mathcal{R}\left(\frac{\partial S}{\partial x}, \frac{\partial S}{\partial y}\right) \frac{\partial S}{\partial y}, \frac{\partial S}{\partial x}\right\rangle dx\, dy \ .$$

Clearly, since $ad - bc \leq \frac{1}{2}(a^2 + d^2 + b^2 + c^2)$ we see that

$$2\left|\int_M \left\langle \mathcal{R}\left(\frac{\partial S}{\partial x}, \frac{\partial S}{\partial y}\right) W(h), W(ih)\right\rangle dx\, dy\right| \leq$$
$$\leq -\int_M (a^2 + b^2 + c^2 + d^2)\left\langle \mathcal{R}\left(\frac{\partial S}{\partial x}, \frac{\partial S}{\partial y}\right) \frac{\partial S}{\partial y}, \frac{\partial S}{\partial x}\right\rangle dx\, dy \ . \tag{6.24}$$

Thus we obtain the inequality

$$Q\left(W(ih), W(ih)\right) + Q\left(W(h), W(h)\right) \leq \frac{1}{2} \sum_{\alpha=1}^{K} \int_M (h \cdot h) g(x)(\nabla_g S^\alpha, \nabla_g S^\alpha) d\mu_g \tag{6.25}$$

which shows that Dirichlet's energy is pluri-subharmonic, but not that it is strictly pluri-subharmonic.

We would like now to show that inequality (6.25) is strict. Our strategy will be as follows: The proof of inequality (6.25) depended on two inequalities, namely (6.22) and (6.24). If either of these were strict, our proof would show that inequality (6.25) is also strict. Our goal is to show that if both inequalities (6.22) and (6.24) are equalities, then $W(h)$ and $W(ih)$ are both zero. In this case strict inequality in (6.25) is trivially satisfied. We begin with inequality (6.24). It follows at once that equality holds iff $a = d$ and $b = -c$. Thus since S is a diffeomorphism we may write:

$$W(ih) = -b\frac{\partial S}{\partial x} + a\frac{\partial S}{\partial y}$$
$$W(h) = a\frac{\partial S}{\partial x} + b\frac{\partial S}{\partial y} \ .$$

Now a simple computation shows that

$$W(h) + iW(ih) = (a - ib)S_z = \bar{\tau}(z)S_z$$
$$W(h) - iW(ih) = (a + ib)S_z = \tau(z)S_z \ ,$$

(6.26)

where $\tau(z) := a + ib$. We now observe that τ is actually a negative differential (i.e. a differential of order -1, $\tau(z)dz^{-1}$. The easiest way to see this is to consider (M, g_0) as isometrically embedded in $\mathbb{R}^K$ with $J_0(x) : T_x M \to T_x M$ the almost complex structure on (M, g_0). Then

$$W(h) - iW(ih) = \tau(z)S_z \ .$$

So

$$\tau(z) = \frac{W(h) - iW(ih)}{S_z} \ .$$

But $S_z dz$ is a differential, and therefore $\tau(z)dz^{-1}$ is a negative differential on (M, g). However such a negative differential is a vector field on (M, g) (one just checks how $\tau(z)$ transforms under conformal changes of coordinates). From this point on we shall simply write $\tau(z)$ for $\tau(z)dz^{-1}$, but will keep in mind that τ is a vector field. Now let us return to inequality (6.22). First observe that since S is harmonic,

$$\nabla_{\frac{\partial}{\partial \bar{z}}} S_z = \nabla_{\frac{\partial}{\partial z}} S_{\bar{z}} = 0$$

(cf. equation (3.2)). Therefore

$$\nabla_{\frac{\partial}{\partial \bar{z}}}\{W(h) - iW(ih)\} = \nabla_{\frac{\partial}{\partial \bar{z}}}(\tau S_z) = \tau_{\bar{z}} S_z + \tau(z)\nabla_{\frac{\partial}{\partial \bar{z}}} S_z = \tau_{\bar{z}} S_z \ .$$

Rewrite (6.21) as

$$2\{Q\left(W(ih), W(ih)\right) + Q\left(W(h), W(h)\right)\} =$$
$$= 4\int_M \langle HS_z, \tau_{\bar{z}} S_z \rangle \ dx \, dy + 4\int_M \langle HS_z, \overline{\tau_{\bar{z}}} S_z \rangle \ dx \, dy \ .$$

If $H = 0$ then $h = 0$ and then $W(h)$ and $W(ih)$ are 0. In this case, strict inequality in (6.25) is satisfied. Now assume that $H \neq 0$. In this case H is a 1-1 tensor corresponding to a holomorphic quadratic differential by raising an index. This implies that both $\operatorname{Re} H$ and $\operatorname{Im} H$ vanish on at most a finite set on M. Consequently in going from (6.21) to (6.22) we have equality iff

$$\tau_z = \kappa(z)H \quad , \quad \tau_{\bar{z}} = \mu(z)\bar{H} \tag{6.27}$$

where κ and μ are real valued functions defined everywhere on M except possibly on a finite set. Since $\operatorname{Re} H \neq 0$, $\operatorname{Im} H \neq 0$ on a dense set, $\tau_{\bar{z}} = 0$ everywhere. Thus τ is a holomorphic vector field on (M, g). Since the only holomorphic vector field is zero, $\tau \equiv 0$; this follows from theorem 2.2.1: Any $\tau \not\equiv 0$ would have a non-trivial flow f_t,

$$\frac{df_t}{dt}(z) = \tau(f_t(z)) \quad , \quad f_0 = id \ .$$

Each such f_t would then be a holomorphic self mapping of $(M, c(g))$ and hence an isometry of (M, g) which is homotopic to the identity f_0. This is ruled out by theorem 2.2.1. Since $\tau \equiv 0$, (6.26) implies that both $W(h)$ and $W(ih)$ must be zero. Therefore we have strict inequality in (6.25). Putting all these facts together we arrive at

Theorem 6.2.6 *Dirichlet's energy $\tilde{E} : T(M) \to \mathbb{R}^+$ on Teichmüller's moduli space is strictly pluri-subharmonic.*

Remark 6.2.5 *Let $\mathbf{L}(W)$ be the Euler-Lagrange operator for harmonic maps from (M, g) to (M, g_0), i.e. in local coordinates*

$$\mathbf{L}(W) = \sum_\ell \left\{ \nabla_{\frac{\partial}{\partial x^\ell}} \nabla_{\frac{\partial}{\partial x^\ell}} W + \mathcal{R}\left(W, \frac{\partial S}{\partial x^\ell}\right) \frac{\partial S}{\partial x^\ell} \right\}$$

where $\nabla_{\frac{\partial}{\partial x^\ell}} = D_{\frac{\partial}{\partial x^\ell}}$ denotes covariant differentiation.
In the proof of the strict pluri-subharmonicity of Dirichlet's energy $\tilde{E}$ on $T(M)$ the vector fields $W(h) \pm W(ih)$ played a prominent role. The reason for this is that equation (6.12) defining $W(h)$ assumes a particularly nice form for $W(h) \pm W(ih)$. A straightforward calculation shows that in local coordinates (6.12) is equivalent to

$$\begin{aligned} \mathbf{L}(W(h) + iW(ih)) &= \nabla_{\frac{\partial}{\partial \bar{z}}} \bar{H} \frac{\partial S}{\partial \bar{z}} \\ \mathbf{L}(W(h) - iW(ih)) &= \nabla_{\frac{\partial}{\partial z}} H \frac{\partial S}{\partial z} \ . \end{aligned} \tag{6.28}$$

This yields yet another proof of the fact that when $g = g_0$, $S = id$, then $W(h) \equiv 0$ (see remark 6.2.4 and (iii) of theorem 3.1.3). Namely if $S = id$, then $\frac{\partial S}{\partial \bar{z}} = 0$, $\frac{\partial S}{\partial z} = 1$, and

$\nabla_{\frac{\partial}{\partial \bar{z}}} H = 0$ since $\bar{H}$ is a 1-1 tensor arising from a holomorphic quadratic differential with one index raised. Thus both $\mathbf{L}(W(h) \pm W(ih))$ vanish, implying that $\mathbf{L}(W(h))$ is zero, and hence $W(h)$ is zero.

Remark 6.2.6 *There are other natural geometric functions which are pluri-subharmonic on $\mathcal{T}(M)$. For example, fix a non-trivial homotopy class of closed curves on (M, g) and let $\ell(g)$ denote the length of the unique closed geodesic in this class. Since $\ell(f^*g) = \ell(g)$ if $f \in \mathcal{D}_0$, ℓ passes to a smooth real valued function on $\mathcal{T}(M)$. Scott Wolpert [124] has shown that ℓ is strictly pluri-subharmonic (actually he proves that ℓ is convex along Weil-Petersson geodesics, which is stronger than being pluri-subharmonic). Unfortunately it is easy to see that ℓ is not proper since geodesics can shrink to zero in length as you "go to infinity" in Teichmüller space. Nevertheless by picking enough such functions he was able to show that a linear combination is proper and therefore yields a strictly pluri-subharmonic (in fact: WP-convex) exhaustion function on M. The geometric idea behind properness is "essentially" the collar lemma in appendix D (although this was not the argument used by Wolpert). Basically as one geodesic shrinks to zero in length, another must go to infinity.*

This result also implies that any two points in $\mathcal{T}(M)$ can be joined by a unique Weil-Petersson geodesic even though $\mathcal{T}(M)$ is not complete in the Weil-Petersson metric. We shall present our own proper Weil-Petersson convex function in section 6.3, thereby re-proving this result of Wolpert's. In particular, our proof indirectly gives a new and concise geometric proof of Wolpert's result (cf. remark 6.3.3)

6.3   Wolf's Form of Dirichlet's Energy on $\mathcal{T}(M)$ is Strictly Weil-Petersson Convex

In his Stanford thesis [123] Michael Wolf introduced another form of Dirichlet's energy on Teichmüller space. Instead of varying the domain metric he varied the range metric. To describe his results, again let

$$\mathsf{E}_G(S) = \mathsf{E}(G, S) = \frac{1}{2} \int_M G_{\alpha\beta}(S) g^{ij} \frac{\partial S^\alpha}{\partial X^i} \frac{\partial S^\beta}{\partial X^j} \, d\mu_g \tag{6.29}$$

denote Dirichlet's energy of a smooth map $S : (M, g) \to (M, G)$ where we now view $g \in \mathcal{M}_{-1}$ as fixed and $G \in \mathcal{M}_{-1}$ as variable, and $G_{\alpha\beta}(S)$ denotes the composition of $G_{\alpha\beta}$ with S. It follows from Appendix B that the unique harmonic diffeomorphism $s(G)$

homotopic to the identity is a smooth C^∞ function of G. We shall henceforth refer to this energy as *Wolf's form of Dirichlet's energy* to distinguish it from the Dirichlet energy $E(g, S)$ introduced by the author. The function E is invariant under the action of $\mathcal{D}$ in the sense that for $f \in \mathcal{D}$

$$\mathsf{E}(G, S) = \mathsf{E}(f^*G, f^{-1}S) \ . \tag{6.30}$$

Since $s(f^*G) = f^{-1}s(G)$ it follows that

$$\mathsf{E}\left(G, s(G)\right) = \mathsf{E}\left(f^*G, s(f^*G)\right) \ , \tag{6.31}$$

and thus E passes to a C^∞ smooth map $\tilde{\mathsf{E}} : T(M) \to \mathbb{R}$, which we call Wolf's form of Dirichlet's energy on Teichmüller space $T(M)$.

Two of Wolf's main results can be stated as follows.

Theorem 6.3.1 (Wolf) $\tilde{\mathsf{E}} : T(M) \to \mathbb{R}^+$ *has only one critical point, namely* $[G] = [g]$. *Moreover the Hessian of* $\tilde{\mathsf{E}}$ *at* $[g] \in T(M)$ *is given by*

$$D^2\tilde{\mathsf{E}}[g](h, k) = \langle h, k \rangle_{WP}$$

for $h, k \in T_{[g]}T(M)$ *and* $\langle \cdot, \cdot \rangle_{WP}$ *the Weil-Petersson inner product*

Remark 6.3.1 *This is analogous to the authors result (iii) in theorem 3.1.3 for* $\tilde{E} : T(M) \to \mathbb{R}^+$.

Let us now write G and g in conformal coordinates, $G_{ij} = \rho\delta_{ij}$, $g_{ij} = \lambda\delta_{ij}$. Then, as in theorem 3.1.3 the expression $\xi(z)dz^2 = 4\rho S_z\overline{S_{\bar{z}}}dz^2$ is a holomorphic quadratic differential on the *fixed* Riemann surface (M, g). Let $[G]$ denote the class of G in $T(M)$.

We then have a second fundamental result of Wolf's [123], namely

Theorem 6.3.2 (Wolf) *The map* $[G] \mapsto \xi(z)dz^2$ *is a diffeomorphism of* $T(M)$ *to the space of holomorphic quadratic differentials on* (M, g).

This last result yields yet another proof (again via harmonic maps) of Teichmüller's result that $T(M) \cong \mathbb{R}^{6\,genus(M)-6}$.

One should also mention that Jost [57] re-derived and somewhat extended Wolf's result. Moreover, in an argument much simpler than that for $\tilde{E}$, Wolf shows

Theorem 6.3.3 (Wolf) $\tilde{\mathsf{E}} : T(M) \to \mathbb{R}^+$ *is a proper map.*

Since the proof of this last theorem is so clearly presented in Jost's book [57] we shall not reproduce it here, save to mention that it again follows from PDE techniques.

In what follows, using the techniques developed in this book, we shall re-derive and extend Wolf's theorem 6.3.1.

There are certain advantages in using Wolf's form of Dirichlet's energy $\tilde{\mathsf{E}}$ as opposed to $\tilde{E}$. One has already been mentioned, namely that the proof of properness is easier. The second is, as we shall shortly see, that $\tilde{\mathsf{E}}$ is strictly Weil-Petersson convex (i.e. strictly convex along Weil-Petersson geodesics) while $\tilde{E}$ is apparently not Weil-Petersson convex (although it is strictly pluri-subharmonic). As in corollary 6.2.3, this implies that $\tilde{\mathsf{E}}$ is also strictly pluri-subharmonic.

The fact that $\tilde{\mathsf{E}}$ is strictly Weil-Petersson convex gives a new proof of Wolpert's results mentioned in Remark 6.2.6 and also a new proof of the Nielsen conjecture which we discuss in the next section. Thus our main goal in this section is to prove

Theorem 6.3.4 $\tilde{\mathsf{E}} : T(M) \to \mathbb{R}^+$ *is strictly Weil-Petersson convex. Moreover* $[g] \in T(M)$ *is the only critical point of* $\tilde{\mathsf{E}}$ *and the Hessian* $D^2\tilde{\mathsf{E}}[g]$ *is given by*

$$D^2\tilde{\mathsf{E}}[g] = \langle \cdot, \cdot \rangle_{WP}$$

We begin working towards our proof.

Let $\sigma(t)$ be a Weil-Petersson geodesic on $T(M)$. Then, as before, $\sigma(t)$ can be lifted to a horizontal path $\tilde{\sigma}$ in $\mathcal{M}_{-1}$. Let $P(G) : T_G\mathcal{M} \to S_2^{TT}(G)$ be the L^2-orthogonal projection from the set S_2 of all symmetric 0-2 tensors (which one identifies with $T_G\mathcal{M}$, $\mathcal{M}$ the manifold of all C^∞ Riemannian metrics) onto $S_2^{TT}(G)$. The condition that $\sigma(t)$ is a Weil-Petersson geodesic can be reformulated as

$$P\left(\tilde{\sigma}(t)\right)\tilde{\sigma}''(t) \equiv 0 \tag{6.32}$$

(cf. equation (6.1)).

Since $\tilde{\sigma}$ is horizontal we have

$$P\left(\tilde{\sigma}(t)\right)\tilde{\sigma}'(t) \equiv \tilde{\sigma}'(t) \tag{6.33}$$

for all t.

Differentiating this last relation we obtain the "lifted" equation of a Weil-Petersson geodesic on $\mathcal{M}_{-1}$ namely

$$\tilde{\sigma}''(t) = [DP\,(\tilde{\sigma}(t))\,\tilde{\sigma}'(t)]\,\tilde{\sigma}'(t) \ , \tag{6.34}$$

D denoting the derivative.

Thus, the lifted geodesic equation depends on the derivative of the projection map P. Remarkably, the fundamental calculation behind the Weil-Petersson convexity of Wolf's form of Dirichlet's energy is the calculation of the derivative of this projection. (Of course, the significance of the derivatives of such projection maps goes back to Gauß' work on surfaces).

Remark 6.3.2 *For most of this course we have not described the L^2-orthogonal complement of $T_G\mathcal{M}_{-1}$ in the space S_2 of symmetric 0-2 tensors. Although we will not need such a description, the beautiful result of Fischer-Marsden [32] is worth mentioning. Let $h \in S_2$, then h can be L^2-orthogonally decomposed as*

$$h = h^{TT} + L_X G + \left\{ \left(\Delta_G f + \frac{1}{2}f\right)G + \mathrm{Hess}\,f\right\}$$

where $h^{TT} + L_X G \in T_G\mathcal{M}_{-1}$, $f : M \to \mathbb{R}$ a smooth function, Δ_G the Laplace-Beltrami operator and Hess f the Hessian of f (the double covariant derivative $\nabla_G\nabla_G f$). Thus all 0-2 tensors in the L^2-orthogonal complement of $T_G\mathcal{M}_{-1}$ are of the form

$$\left\{ \left(\Delta_G f + \frac{1}{2}f\right)G + \mathrm{Hess}\,f\right\} \ .$$

We now prove

Theorem 6.3.5 *Let $h \in S_2^{TT}(G)$, then*

$$DP(G)[h]h = \lambda G + L_Z G$$

where

$$\lambda = \frac{1}{2}(h \cdot h) = \langle D\pi h, D\pi h\rangle_{WP} \ ,$$

and the trace $\mathrm{tr}_G(L_Z G) = 0$. Here, Z is given by theorem 1.4.2, $\pi : \mathcal{M}_{-1} \to \mathcal{T}(M)$ is the principal bundle projection (cf. section 2.5), and $\cdot$ denotes the L^2-inner product density on 0-2 tensors (with respect to the metric G).

Since $\delta_G Z = \frac{1}{2}\text{tr}_G(L_Z G)$, $L_Z G$ above is tangent to the space of volume preserving diffeomorphisms (cf. theorem 5.3.4).

PROOF: Again δ_G is the divergence, $\alpha_G : \mathcal{X}(M) \to S_2$ is the Lie derivative $X \mapsto L_X G$, and $\Pi(G) : S_2 \to S_2^\tau(G)$ denotes the L^2-orthogonal projection of the symmetric 0-2 tensors onto their trace free parts (cf. p. 85 for the analogous projection on 1-1 tensors).

We know that, given G, any $h \in S_2$ can be L^2-orthogonally decomposed as (theorem 1.4.2)

$$h = h^0 + L_X G$$

where $\delta_G h^0 = 0$.

In terms of the operators α_G, δ_G, we can write

$$h^0 = \left\{ h - \alpha_G(\delta_G\alpha_G)^{-1}\delta_G h \right\}$$

and

$$L_X G = \alpha_G(\delta_G\alpha_G)^{-1}\delta_G h \ .$$

Here, we are using the fact that $\delta_G\alpha_G = -\delta_G\delta_G^*$ is an isomorphism (cf. remark 1.4.1).

It follows that

$$
\begin{aligned}
P(G)h &= \Pi(G)\left\{ h - \alpha_G(\delta_G\alpha_G)^{-1}\delta_G h \right\} \\
&= \Pi(G)h - \Pi(G)\alpha_G(\delta_G\alpha_G)^{-1}\delta_G h \ .
\end{aligned}
\tag{6.35}
$$

Lemma 6.3.6 *If* $h \in S_2^{TT}(G)$, *the derivative of the map* $G \mapsto \Pi(G)\alpha_G(\delta_G\alpha_G)^{-1}\delta_G h$ *in the direction* h *is*

$$h \mapsto \Pi(G)\alpha_G(\delta_G\alpha_G)^{-1}(D_G\delta_G)(h)h \ ,$$

where $D_G\delta_G$ *denotes the derivative of* δ_G *w.r.t.* G.

PROOF: A straightforward calculation.   ∎

Lemma 6.3.7 $k = \alpha_G(\delta_G\alpha_G)^{-1}(D_G\delta_G)(h)h$ *is trace free.*

PROOF: By equation (5.15) it follows that

$$(D_G\delta_G)(h)h = -\frac{1}{2}*d\mu \ ,$$

μ a smooth real valued function on M and $*d\mu$ the Hodge dual of $d\mu$. Thus

$$\delta_G(D_G\delta_G)(h)h = -\frac{1}{2}\delta_G(*d\mu) = 0$$

(see also lemma E.2).

Thus, if $X = (D_G\delta_G)(h)h$, then $k = \alpha_G(\delta_G\alpha_G)^{-1}X$, where $\delta_G X = 0$. But $\delta_G k = X$. Thus

$$\delta_G\delta_G k = 0 \ .$$

Since k is a Lie derivative it is automatically tangent to $\mathcal{M}_{-1}$ at G. Thus (cf. theorem 2.5.2 and formula (1.5)) k satisfies the PDE

$$0 = DR(G)k = -\Delta_G(\mathrm{tr}_G k) + \frac{1}{2}\mathrm{tr}_G k + \delta_G\delta_G k \ .$$

Hence

$$-\Delta_G(\mathrm{tr}_G k) + \frac{1}{2}\mathrm{tr}_G k = 0 \ ,$$

and this immediately implies, as before, that $\mathrm{tr}_G k = 0$, proving 6.3.7. ∎

This allows us to strengthen lemma 6.3.6 to

Lemma 6.3.8 *If $h \in S_2^{TT}(G)$, the derivative of the map $G \mapsto \Pi(G)\alpha_G(\delta_G\alpha_G)^{-1}\delta_G h$ in the direction h is*

$$h \mapsto \alpha_G(\delta_G\alpha_G)^{-1}X = \alpha_G Y$$

where $X = (D_G\delta_G)(h)h$, $\delta_G X = 0$, $\delta_G Y = \frac{1}{2}\mathrm{tr}_G(\alpha_G Y) = \frac{1}{2}\mathrm{tr}_G k = 0$. ∎

Lemma 6.3.9 *If $\mathrm{tr}_G h = 0$, the derivative*

$$D\Pi(G)[h]h = \lambda G$$

where $\lambda = \frac{1}{2}(h \cdot h) = \langle D\pi h, D\pi h \rangle_{WP}$

PROOF:

$$\Pi(G)h = h - \frac{1}{2}(\mathrm{tr}_G h)G$$

Thus

$$D\Pi(G)[h]h = -\frac{1}{2}(\mathrm{tr}_G h)h - \frac{1}{2}D_G(\mathrm{tr}_G h)[h]G \ ,$$

D_G denoting derivative w.r.t. G.

Now

$$\mathrm{tr}_G h = G^{ij} h_{ij} \ ,$$

hence

$$D_G(\mathrm{tr}_G h)[h] = -h^{ij} h_{ij} = -h \cdot h \ .$$

Thus $D\Pi(G)[h]h = \frac{1}{2}(h \cdot h)\, G$.   ∎

Putting lemmas 6.3.8 and 6.3.9 together we obtain the proof of theorem 6.3.5.   ∎

We are now ready to proceed further with the proof of the Weil-Petersson convexity of $\tilde{\mathsf{E}}$. As we have done before, we will denote by $s(G)$ the unique smooth harmonic map from (M,g) to (M,G) which is homotopic to the identity. (We already used this notation in the discussion of the Earle-Eells section of $(\pi, \mathcal{M}_{-1}, \mathcal{T}(M))$, cf. p. 81).

Let $\sigma(t)$ be a Weil-Petersson geodesic on $\mathcal{T}(M)$ and $\tilde{\sigma}(t)$ its horizontal lift to $\mathcal{M}_{-1}$. For simplicity let $G^t := \tilde{\sigma}(t)$. Wolf's form of Dirichlet's energy along $\sigma(t)$ can then be expressed as

$$\begin{aligned}
\tilde{\mathsf{E}}(\sigma(t)) \; &= \; \mathsf{E}\left(G^t, s(G^t)\right) \\
&= \; \frac{1}{2} \int\limits_M G^t_{\alpha\beta}(s) g^{ij} \frac{\partial s^\alpha}{\partial x^i} \frac{\partial s^\beta}{\partial x^j} d\mu_g \ ,
\end{aligned}$$

and in conformal coordinates on (M,g), this is

$$\frac{1}{2} \int G^t_{\alpha\beta}(s) \frac{\partial s^\alpha}{\partial x^i} \frac{\partial s^\beta}{\partial x^i} dx \, dy \ ,$$

where $(x^1, x^2) = (x, y)$.

Therefore if $\frac{d}{dt}\tilde{\sigma}(t) = H^t$, H^t a 0-2 tensor tangent to $\mathcal{M}_{-1}$ at G,

$$\begin{aligned}
\frac{d}{dt}\tilde{\mathsf{E}}(\sigma(t)) \; = \; & \frac{1}{2} \int H_{\alpha\beta}(s) \frac{\partial s^\alpha}{\partial x^i} \frac{\partial s^\beta}{\partial x^i} dx \, dy \\
& + \frac{1}{2} \int G_{\alpha\beta,j}(s) W^j \frac{\partial s^\alpha}{\partial x^i} \frac{\partial s^\beta}{\partial x^i} dx \, dy \qquad\qquad (6.36) \\
& + \int G_{\alpha\beta}(s) \frac{\partial W^\alpha}{\partial x^i} \frac{\partial s^\beta}{\partial x^i} dx \, dy
\end{aligned}$$

Here we have suppressed the t dependence of $H_{\alpha\beta}$, $G_{\alpha\beta}$, and s, and we use the notation that $G_{\alpha\beta,j}$ is the derivative of $G_{\alpha\beta}$ w.r.t. the jth variable. Moreover $W = Ds(G)H$ is a vector field over s.

Lemma 6.3.10 *Let* E_G *be the Dirichlet energy for harmonic maps (G and g fixed, then* $E_G = E_g$*). Then the second variation of* E_G *w.r.t. mappings* $s : (M, g) \to (M, G)$ *in the direction* W *is*

$$D^2 E_G(W, W) = -\frac{1}{2} \int H_{\alpha\beta,j}(s) W^j \frac{\partial s^\alpha}{\partial x^i} \frac{\partial s^\beta}{\partial x^i} dx\, dy - \int H_{\alpha\beta}(s) \frac{\partial W^\alpha}{\partial x^i} \frac{\partial s^\beta}{\partial x^i} dx\, dy \ .$$

PROOF: Since s is harmonic

$$0 \equiv \frac{1}{2} \int G_{\alpha\beta,j}(s) W^j \frac{\partial s^\alpha}{\partial x^i} \frac{\partial s^\beta}{\partial x^i} dx\, dy + \int G_{\alpha\beta}(s) \frac{\partial W^\alpha}{\partial x^i} \frac{\partial s^\beta}{\partial x^i} dx\, dy \ . \qquad (6.37)$$

Differentiate this identity w.r.t. t and use the definition of the second variation. The reader should compare this calculation with lemma 6.2.5 where the domain metric is varied. ∎

Lemma 6.3.11 *Let* $H^{tt}_{\alpha\beta} := \frac{d^2}{dt^2} \tilde\sigma(t)$*. Then*

$$\begin{aligned}
\frac{d^2}{dt^2} \tilde E(\sigma(t)) &= \frac{1}{2} \int H^{tt}_{\alpha\beta}(s) \frac{\partial s^\alpha}{\partial x^i} \frac{\partial s^\beta}{\partial x^i} dx\, dy + \frac{1}{2} \int H_{\alpha\beta,j}(s) W^j \frac{\partial s^\alpha}{\partial x^i} \frac{\partial s^\beta}{\partial x^i} dx\, dy \\
&\quad + \int H_{\alpha\beta}(s) \frac{\partial W^\alpha}{\partial x^i} \frac{\partial s^\beta}{\partial x^i} dx\, dy
\end{aligned}$$

PROOF: Differentiate (6.36) w.r.t. t and apply (6.37). ∎

We now come to the fundamental formula for the second variation of Wolf's form of Dirichlet's energy along Weil-Petersson geodesics.

Lemma 6.3.12

$$\frac{d^2 \tilde E}{dt^2}(\sigma(t)) = \frac{1}{4} \int (H \cdot H)\, G_{\alpha\beta}(s) g^{ij} \frac{\partial s^\alpha}{\partial x^i} \frac{\partial s^\beta}{\partial x^j} d\mu_g - D^2 E_G(W, W) \ , \qquad (6.38)$$

$W = DS(g)H$.

PROOF: Since $\sigma(t)$ is a Weil-Petersson geodesic, we have by (6.34)

$$H^{tt} = DP(\tilde\sigma(t))\, [\tilde\sigma'(t)]\, \tilde\sigma'(t) \ .$$

By theorem 6.3.5

$$DP(G)[H](H) = \lambda G + L_Z G$$

where $\lambda = \frac{1}{2}(H \cdot H)$. Applying lemmas 6.3.10 and 6.3.11 we see that

$$
\begin{aligned}
\frac{d^2\check{\mathsf{E}}}{dt^2}(\sigma(t)) = {} & \frac{1}{4}\int (H \cdot H)\, G_{\alpha\beta}(s)\frac{\partial s^\alpha}{\partial x^i}\frac{\partial s^\beta}{\partial x^i}dx\,dy \\
& + \frac{1}{2}\int (L_Z G)_{\alpha\beta}\frac{\partial s^\alpha}{\partial x^i}\frac{\partial s^\beta}{\partial x^i}dx\,dy \\
& - D^2\mathsf{E}_G(W, W)
\end{aligned} \tag{6.39}
$$

We claim that the second term on the right side of the above inequality must vanish. This is a consequence of the group invariance of Dirichlet's energy (cf. (6.31)). To see this, differentiate the identity

$$
\mathsf{E}(G, s(G)) = \mathsf{E}(f_t^*G, s(f_t^*G)) , \tag{6.40}
$$

f_t a family of diffeomorphisms, $f_0 = id$, $\frac{df}{dt}\big|_{t=0} = Z$. Then, since $\frac{d}{dt}f_t^*G\big|_{t=0} = L_Z G$ and the left side of (6.40) does not depend on t, we obtain

$$
0 \equiv \frac{1}{2}\int (L_Z G)_{\alpha\beta}\frac{\partial s^\alpha}{\partial x^i}\frac{\partial s^\beta}{\partial x^i}dx\,dy + D\mathsf{E}_G(s)(W)
$$

where $W = \frac{d}{dt}s(f_t^*G)\big|_{t=0}$. Since s is harmonic $D\mathsf{E}_G(s) \equiv 0$, verifying the claim. This proves lemma 6.3.12. ∎

Remark 6.3.3 *According to corollary 6.2.3, formula 6.38 yields the Levi-form of Wolf's form of Dirichlet's energy in some complex coordinate system, and by remark 6.2.3 the Levi-form (up to a scalar multiple) of Wolf's form of Dirichlet's energy in Abresch-Fischer coordinates. Since there is also a very minor difference between Wolf's form of Dirichlet's energy and geodesic length functionals studied by Wolpert [124] we observe that in all three cases of geometric functionals on Teichmüller's moduli space (Dirichlet's energy, Wolf's form of Dirichlet's energy and geodesic length functionals) the Levi-forms have the same structure, namely the difference of the integral of Dirichlet's energy density weighted by the Weil-Petersson density and the sum of two second variations of Dirichlet's energy. In the present situation*

$$
\begin{aligned}
\frac{D^2\check{\mathsf{E}}}{\partial z^\gamma \bar{z}^{\hat{\jmath}}}([G]) = {} & \frac{1}{2}\int_M (H \cdot H)\, G_{\alpha\beta}(s)g^{ij}\frac{\partial s^\alpha}{\partial x^i}\frac{\partial s^\beta}{\partial x^j}dx\,dy \\
& - D^2\mathsf{E}_G(s)(W(H), W(H)) - D^2\mathsf{E}_G(s)(W(iH), W(iH))
\end{aligned} \tag{6.41}
$$

where $W(H) = D_G s(G)H$. For geodesics, $G_{\alpha\beta}(s)g^{ij}\frac{\partial s^\alpha}{\partial x^i}\frac{\partial s^\beta}{\partial x^j}d\mu_g$ is replaced by the geodesic energy density and $D^2\mathsf{E}_G$ by the second variation of geodesic energy.

We would now like to complete our proof that Wolf's form of Dirichlet's energy is Weil-Petersson convex. In the interest of brevity, we point out, as the readers may verify for themselves, that if $H \in S_2^{TT}(G)$, then

$$D^2 E_G(s)(W(H), W(H)) = D^2 E_G(s)(W(iH), W(iH)) \qquad (6.42)$$

This follows directly from lemma 6.3.10 and a straightforward computation. In view of (6.41), (6.42) says that if Wolf's form of Dirichlet's energy is pluri-subharmonic it must also be Weil-Petersson convex. In fact, (6.42) is strong evidence for the Weil-Petersson convexity of $\tilde{E}$. This also yields a simplification of (6.41) for the Levi-form of $\tilde{E}$, namely

$$\frac{\partial^2 \tilde{E}}{\partial z^\gamma \partial \bar{z}^\beta}([G]) = \frac{1}{2} \int (H \cdot H) \, G_{\alpha\beta}(s) g^{ij} \frac{\partial s^\alpha}{\partial x^i} \frac{\partial s^\beta}{\partial x^j} d\mu_g - 2D^2 E_G(s)(W(H), W(H))$$

By lemma 6.3.12, in conformal coordinates, $g_{ij} = \lambda \delta_{ij}$,

$$\frac{d^2 \tilde{E}}{dt^2}(\sigma(t)) = \frac{1}{4} \int (H \cdot H) \, G_{\alpha\beta}(s) \frac{\partial s^\alpha}{\partial x^i} \frac{\partial s^\beta}{\partial x^i} dx \, dy - D^2 E_G(W, W)$$

where $W = Ds(G)H$, $H \in S_2^{TT}(G)$.

Consequently, in order to establish the Weil-Petersson convexity of $\tilde{E}$ it suffices to show that for $H \neq 0$,

$$D^2 E_G(W, W) < \frac{1}{4} \int (H \cdot H) \, G_{\alpha\beta}(s) \frac{\partial s^\alpha}{\partial x^i} \frac{\partial s^\beta}{\partial x^i} dx \, dy \ . \qquad (6.43)$$

From lemma 6.3.10 we have a formula for $D^2 E_G$, namely

$$
\begin{aligned}
-D^2 E_G(W, W) &= \frac{1}{2} \int H_{\alpha\beta,j}(s) W^j \frac{\partial s^\alpha}{\partial x^i} \frac{\partial s^\beta}{\partial x^i} dx \, dy + \int H_{\alpha\beta}(s) \frac{\partial W^\alpha}{\partial x^i} \frac{\partial s^\beta}{\partial x^i} dx \, dy \\
&= \frac{1}{2} \int H_{\alpha\beta,j}(s) W^j \frac{\partial s^\alpha}{\partial x^i} \frac{\partial s^\beta}{\partial x^i} dx \, dy + \frac{1}{2} \int H_{\alpha\beta}(s) \frac{\partial W^\alpha}{\partial x^i} \frac{\partial s^\beta}{\partial x^i} dx \, dy \\
&\quad + \frac{1}{2} \int H_{\alpha\beta}(s) \frac{\partial W^\alpha}{\partial x^i} \frac{\partial s^\beta}{\partial x^i} dx \, dy \ .
\end{aligned}
$$

Lemma 6.3.13 *For $H \in S_2^{TT}(G)$*

$$
\begin{aligned}
\frac{1}{2} \int H_{\alpha\beta,j}(s) W^j \frac{\partial s^\alpha}{\partial x^i} \frac{\partial s^\beta}{\partial x^i} dx \, dy &+ \frac{1}{2} \int H_{\alpha\beta}(s) \frac{\partial W^\alpha}{\partial x^i} \frac{\partial s^\beta}{\partial x^i} dx \, dy = \\
&= -\frac{1}{2} \int H_{\alpha\beta}(s) W^\alpha \frac{\partial^2 s^\beta}{\partial x^i \partial x^i} dx \, dy \ .
\end{aligned}
\qquad (6.44)
$$

PROOF: Integrating the second term on the left by parts we see that the left hand side of (6.44) is equal to

$$\frac{1}{2}\int H_{\alpha\beta,j}(s)W^j\frac{\partial s^\alpha}{\partial x^i}\frac{\partial s^\beta}{\partial x^i}dx\,dy - \frac{1}{2}\int H_{\alpha\beta,j}(s)W^\alpha\frac{\partial s^j}{\partial x^i}\frac{\partial s^\beta}{\partial x^i}dx\,dy$$

$$-\int H_{\alpha\beta}(s)W^\alpha\frac{\partial^2 s^\beta}{\partial x^i\partial x^i}dx\,dy \ . \tag{6.45}$$

From the fact that $H\in S_2^{TT}(G)$, the first two terms cancel yielding the lemma. ■

Lemma 6.3.14

$$-\frac{1}{2}\int H_{\alpha\beta}(s)W^\alpha\frac{\partial^2 s^\beta}{\partial x^i\partial x^i}dx\,dy + \frac{1}{2}\int H_{\alpha\beta}(s)\frac{\partial W^\alpha}{\partial x^i}\frac{\partial s^\beta}{\partial x^i}dx\,dy =$$
$$= \frac{1}{2}\int H_{\alpha\beta}(s)\left(\nabla_{\frac{\partial}{\partial x^i}}W^\alpha\right)\frac{\partial s^\beta}{\partial x^i}dx\,dy$$

PROOF: Since s is harmonic

$$\frac{\partial^2 s^\beta}{\partial x^i\partial x^i} = -\Gamma^\beta_{\gamma\mu}(s)\frac{\partial s^\gamma}{\partial x^i}\frac{\partial s^\mu}{\partial x^i} \ .$$

Thus

$$-\frac{1}{2}\int H_{\alpha\beta}(s)W^\alpha\frac{\partial^2 S^\beta}{\partial x^i\partial x^i}dx\,dy + \frac{1}{2}\int H_{\alpha\beta}(s)\frac{\partial W^\alpha}{\partial x^i}\frac{\partial s^\beta}{\partial x^i}dx\,dy$$

$$= \frac{1}{2}\int H_{\alpha\beta}(s)W^\alpha\Gamma^\beta_{\gamma\mu}(s)\frac{\partial s^\gamma}{\partial x^i}\frac{\partial s^\mu}{\partial x^j}dx\,dy + \frac{1}{2}\int H_{\alpha\beta}(s)\frac{\partial W^\alpha}{\partial x^i}\frac{\partial s^\beta}{\partial x^i}dx\,dy \ . \tag{6.46}$$

Now

$$\left(\nabla_{\frac{\partial}{\partial x^i}}W\right)^\alpha = \frac{\partial W^\alpha}{\partial x^i} + \Gamma^\alpha_{\epsilon\nu}(s)W^\epsilon\frac{\partial s^\nu}{\partial x^i} \ .$$

Thus (6.46) is equal to

$$\frac{1}{2}\int H_{\alpha\beta}(s)W^\alpha\Gamma^\beta_{\gamma\mu}(s)\frac{\partial s^\gamma}{\partial x^i}\frac{\partial s^\mu}{\partial x^i}dx\,dy \quad - \quad \frac{1}{2}\int H_{\alpha\beta}(s)W^\epsilon\Gamma^\alpha_{\epsilon\nu}(s)\frac{\partial s^\nu}{\partial x^i}\frac{\partial s^\beta}{\partial x^i}dx\,dy$$

$$+ \frac{1}{2}\int H_{\alpha\beta}(s)\left(\nabla_{\frac{\partial}{\partial x^i}}W\right)^\alpha\frac{\partial s^\beta}{\partial x^i}dx\,dy \ .$$

We claim that the sum of the first two terms above is zero, and clearly this fact proves the lemma. This claim follows from the symmetry of H, the fact that H is trace free, and that the Christoffel symbols in G-conformal coordinates satisfy $\Gamma^2_{12} = \Gamma^2_{21} = \Gamma^1_{11} = -\Gamma^1_{22}$ and $\Gamma^1_{12} = \Gamma^1_{21} = \Gamma^2_{22} = -\Gamma^2_{11}$. ■

Putting lemmas 6.3.13 and 6.3.14 together we conclude

Lemma 6.3.15

$$-D^2 \mathsf{E}_G(W, W) = \frac{1}{2} \int H_{\alpha\beta}(s) \left(\nabla_{\frac{\partial}{\partial x^i}} W \right)^\alpha \frac{\partial s^\beta}{\partial x^i} dx \, dy$$

We are now ready to establish inequality (6.43) and thus the strict Weil-Petersson convexity of Wolf's Dirichlet energy.

Lemma 6.3.16 *In conformal coordinates $g_{ij} = \lambda \delta_{ij}$, and for $H \neq 0$,*

$$D^2 \mathsf{E}_G(W, W) < \frac{1}{4} \int (H \cdot H) \, G_{\alpha\beta} \frac{\partial s^\alpha}{\partial x^i} \frac{\partial s^\beta}{\partial x^i} dx \, dy$$

PROOF: By the previous lemma

$$
\begin{aligned}
D^2 \mathsf{E}_G(W, W) = \; & -\frac{1}{2} \int \left\{ H_{11} \left(\nabla_{\frac{\partial}{\partial x^i}} W \right)^1 + H_{12} \left(\nabla_{\frac{\partial}{\partial x^i}} W \right)^2 \right\} \frac{\partial s^1}{\partial x^i} dx \, dy \\
& -\frac{1}{2} \int \left\{ -H_{11} \left(\nabla_{\frac{\partial}{\partial x^i}} W \right)^2 + H_{12} \left(\nabla_{\frac{\partial}{\partial x^i}} W \right)^1 \right\} \frac{\partial s^2}{\partial x^i} dx \, dy \; .
\end{aligned}
$$

Writing $G_{\alpha\beta} = \rho \delta_{\alpha\beta}$ and applying the inequality $AB \leq (A^2 + B^2)/2$, we find that

$$
\begin{aligned}
D^2 \mathsf{E}_G(W, W) \leq \; & \frac{1}{8} \sum_{i,j} \int \left\{ (H_1^1)^2 + (H_2^1)^2 \right\} \rho \left\{ \frac{\partial s^j}{\partial x^i} \right\}^2 dx \, dy \\
& + \frac{1}{2} \sum_{i,j} \int \rho \left\{ \left(\nabla_{\frac{\partial}{\partial x^i}} W \right)^j \right\}^2 dx \, dy \\
= \; & \frac{1}{16} \int (H \cdot H) \, G_{\alpha\beta}(s) g^{ij} \frac{\partial s^\alpha}{\partial x^i} \frac{\partial s^\beta}{\partial x^j} d\mu_g \\
& + \frac{1}{2} \sum_i \int \left\| \nabla_{\frac{\partial}{\partial x^i}} W \right\|^2 dx \, dy \; .
\end{aligned}
$$

But (cf. formula (6.9)) since

$$\sum_i \int \left\| \nabla_{\frac{\partial}{\partial x^i}} W \right\|^2 dx \, dy \leq D^2 \mathsf{E}_G(W, W)$$

we obtain the inequality

$$D^2 \mathsf{E}_G(W, W) \leq \frac{1}{8} \int (H \cdot H) \, G_{\alpha\beta}(s) g^{ij} \frac{\partial s^\alpha}{\partial x^i} \frac{\partial s^\beta}{\partial x^j} d\mu_g \; ,$$

which, if $H \neq 0$, is

$$< \frac{1}{4} \int (H \cdot H) \, G_{\alpha\beta}(s) g^{ij} \frac{\partial s^\alpha}{\partial x^i} \frac{\partial s^\beta}{\partial x^j} d\mu_g \; ,$$

which is the lemma in general coordinates $\{g_{ij}\}$. This establishes the Weil-Petersson convexity part of theorem 6.3.4.

We now argue that $\tilde{\mathbb{E}} : T(M) \to \mathbb{R}^+$ (as with $\tilde{E} : T(M) \to \mathbb{R}^+$) can have only one critical point (clearly $[G] = [g]$). This follows immediately from

Lemma 6.3.17 *Let $\tilde{\mathbb{E}} : T(M) \to \mathbb{R}^+$ be any proper smooth map which is strictly Weil-Petersson convex. Then $\tilde{\mathbb{E}}$ has a unique minimum, and this is the only critical point.*

PROOF: That $\tilde{\mathbb{E}}$ has a minimum follows immediately from properness. Moreover by strict Weil-Petersson convexity it follows that all critical points are non-degenerate minima. Thus if $\tilde{\mathbb{E}}$ should have more than one critical point, then by elementary Morse-theoretic arguments [64] there must be a third critical point which is not a minimum. In the mathematical literature this is also called a "saddle point principle"; this principle was in fact first used by Morse, Shiffman and Tompkins in the theory of minimal surfaces. ∎

Intuitively the justification of the saddle point principle is clear: If $\tilde{\mathbb{E}} : \mathbb{R}^N \to \mathbb{R}^+$ has, say, two non-degenerate minima p_1 and p_2, $\tilde{\mathbb{E}}(p_1) \geq \tilde{\mathbb{E}}(p_2)$, and no other critical points, then the level sets $\tilde{\mathbb{E}}^{-1}[0, a]$ will be disconnected for all $a > \tilde{\mathbb{E}}(p_2)$. This, of course, is impossible, and the reason why is shown schematically in figure 6.1 in the case $N = 2$. The saddle point principle also follows from the deeper Morse inequalities [64].

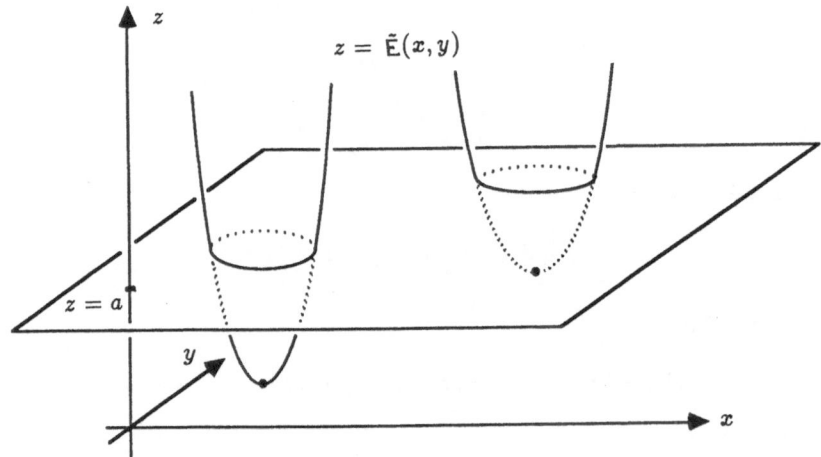

Figure 6.1: The impossibility of a proper C^2 function defined on all of $\mathbb{R}^2$ with only two non-degenerate minima.

Thus we have established the second part of theorem 6.3.4. For the last part we follow a line of reasoning exactly analogous to lemma 3.1.5, and therefore we shall be somewhat brief.

Lemma 6.3.18 *Let $s(G) : (M, g) \to (M, G)$ be the unique harmonic map homotopic to the identity. Clearly for $g = G$, $s(G) = id$. Then if $W = Ds(id)H$ with $H \in S_2^{TT}(G)$, $W = 0$.*

PROOF: The equation for $s(G)$ is

$$\mathbf{L}_G(s) := \frac{1}{\sqrt{g}} \frac{\partial}{\partial x^i} g^{ij} \sqrt{g} \frac{\partial s^\alpha}{\partial x^j} + \Gamma^\alpha_{\gamma\beta}(s) \frac{\partial s^\gamma}{\partial x^i} \frac{\partial s^\beta}{\partial x^j} g^{ij} = 0 . \tag{6.47}$$

(It's not the same L here as in remark 6.2.5.) Differentiating the operator $\mathbf{L}_G$ w.r.t. G and integrating against $W = Ds(id)H$ we obtain

$$D^2 \mathbf{E}_G(W, W) = \int \varphi^\alpha W^\alpha d\mu_g$$

where φ, in conformal coordinates $G_{\alpha\beta} = \rho\delta_{\alpha\beta}$, $g_{ij} = \lambda\delta_{ij}$, can be written as

$$\begin{aligned}
\varphi^\alpha &= \frac{1}{2} H^{\alpha k} \left\{ \frac{\partial \rho}{\partial x^\beta} \delta_{\gamma k} + \frac{\partial \rho}{\partial x^\gamma} \delta_{\beta k} - \frac{\partial \rho}{\partial x^k} \delta_{\gamma\beta} \right\} \frac{1}{\lambda} \frac{\partial s^\gamma}{\partial x^i} \frac{\partial s^\beta}{\partial x^i} \\
&+ \frac{\rho}{2} \delta^{\alpha k} \left\{ \frac{\partial H_{\gamma k}}{\partial x^\beta} + \frac{\partial H_{\beta k}}{\partial x^\gamma} - \frac{\partial H_{\gamma\beta}}{\partial x^k} \right\} \frac{1}{\lambda} \frac{\partial s^\gamma}{\partial x^i} \frac{\partial s^\beta}{\partial x^i} ,
\end{aligned}$$

because

$$\Gamma^\alpha_{\gamma\beta} = \frac{1}{2} G^{\alpha k} \left\{ \frac{\partial G_{\gamma k}}{\partial x^\beta} + \frac{\partial G_{\beta k}}{\partial x^\gamma} - \frac{\partial G_{\gamma\beta}}{\partial x^k} \right\} .$$

Then, since $\frac{\partial s^\gamma}{\partial x^i} = \delta_{\gamma i}$,

$$\varphi^\alpha = \frac{1}{2\lambda} \left\{ \frac{\partial \rho}{\partial x^i} \delta_{ik} + \frac{\partial \rho}{\partial x^i} \delta_{ik} - \frac{\partial \rho}{\partial x^k} \delta_{ii} \right\} + \frac{\rho}{2\lambda} \left\{ \frac{\partial H_{ik}}{\partial x^i} + \frac{\partial H_{ik}}{\partial x^i} - \frac{\partial H_{ii}}{\partial x^k} \right\} .$$

The first bracket is obviously zero, and the second bracket is zero since $H \in S_2^{TT}(G)$. Thus $D^2 \mathbf{E}_G(W, W) = 0$, and hence $W = 0$. This concludes 6.3.18. ■

Back to the last part of theorem 6.3.4. We know that for $\tilde{\mathbf{E}} : T(M) \to \mathbb{R}^+$, the unique critical point is $[G] = [g]$. Let $H \in S_2^{TT}(g)$ and $\tilde{H} = D\pi(H) \in T_{[g]}T(M)$. Then if σ is a geodesic in $T(M)$ with $\sigma(0) = [g]$ and $\sigma'(0) = \tilde{H}$, the Hessian

$$D^2\tilde{\mathbf{E}}[g](\tilde{H}, \tilde{H}) = \frac{d^2\tilde{\mathbf{E}}}{dt^2}(\sigma(t)) ,$$

which by lemma 6.3.12 and the last lemma equals

$$\frac{1}{4}\int (H\cdot H)\, G_{\alpha\beta}g^{ij}\frac{\partial s^\alpha}{\partial x^i}\frac{\partial s^\beta}{\partial x^j}d\mu_g \ .$$

For $g = G$ and $s = id$, $G_{\alpha\beta}g^{ij}\frac{\partial s^\alpha}{\partial x^i}\frac{\partial s^\beta}{\partial x^j} = 2$.

Thus

$$D^2\tilde{\mathsf{E}}[g](\tilde{H},\tilde{H}) = \frac{1}{2}\int (H\cdot H)\, d\mu_g = \left\langle \tilde{H},\tilde{H}\right\rangle_{WP} \ .$$

We also recover the following result of Wolpert's from Weil-Petersson convexity of $\tilde{\mathsf{E}}$, namely

Theorem 6.3.19 *Any two points in Teichmüller space can be joined by a (unique) Weil-Petersson geodesic.*

PROOF: Since $\tilde{\mathsf{E}}$ is geodesically convex and proper, the non-empty compact balls $\tilde{\mathsf{E}}^{-1}[0,R]$ have the property that any two points can be joined by a unique geodesic. The result follows, because $T(M) = \bigcup_R \tilde{\mathsf{E}}^{-1}[0,R]$.   ∎

6.4   The Nielsen Realization Problem

The *Nielsen Realization Problem*, now a half century old, has attracted the attention of many mathematicians in diverse fields, and as a result has had an interesting, if not tumultuous, history. What is the problem?

Fix a surface M, genus $M > 1$. Basically one asks whether any finite subgroup $\hat{\mathcal{G}} \subset \mathcal{D}/\mathcal{D}_0$ can be "realized" as a subgroup $\mathcal{G} \subset \mathcal{D}$.

The problem was first raised, in essentially this form, by Nielsen and solved by him in 1942 [86] in the case that $\hat{\mathcal{G}}$ is cyclic. In 1948 and 1950 Fenchel [30], [31] reproved Nielsen's result and extended it, using the Smith fixed point theorem [101] (namely that every periodic map $f : \mathbb{R}^n \to \mathbb{R}^n$ of prime order has a fixed point) to finite solvable groups $\hat{\mathcal{G}}$.

In this work Fenchel also reformulates the problem in terms of Teichmüller theory as follows: If the action of $\hat{\mathcal{G}} \subset \mathcal{D}/\mathcal{D}_0$ on $T(M)$ has a fixed point then $\hat{\mathcal{G}}$ can be realized as a subgroup $\mathcal{G}$ of $\mathcal{D}$.

To see why this is so, let us consider, for simplicity of exposition, the case where $\hat{\mathcal{G}}$ is finite cyclic. Then $\hat{\mathcal{G}} = \{[id], [f], [f^2], \ldots, [f^{p-1}]\}$, $f \in \mathcal{D}$. Suppose the action of $\hat{\mathcal{G}}$ on $\mathcal{T}(M)$ has a fixed point $[g] \in \mathcal{M}_{-1}/\mathcal{D}_0$. This says that $[f^*g] = [g]$ or that the class $[f]$ is representable as a holomorphic map, say f, with respect to the complex structure $c(g)$ associated to $[g]$. Therefore $[f]^p = id$ means that f^p is homotopic to the identity, by theorem 2.2.1, this implies that $f^p = id$, as a map. Therefore the cyclic group $\{id, f, f^2, \ldots, f^{p-1}\} = \mathcal{G}$ "realizes" $\hat{\mathcal{G}}$ in $\mathcal{D}$. The case for general $\mathcal{G}$ follows after observing that every relation in $\mathcal{G}$ can be realized as a relation on a set of maps.

Let us continue with our history. In 1959, Sol Kravetz [61], using Fenchel's reformulation of the problem gave a positive answer for the general case. His proof used curvature properties of the Teichmüller metric on $\mathcal{T}(M)$ (this Finsler metric is not discussed in these notes). In 1962 Macbeath again using Teichmüller theory and Kravetz's curvature result gives another proof [70].

In 1971 Zieschang [128] showed that a finite torsion free extension of a surface group $\hat{\mathcal{G}}$ is again a surface group, and at the same time gave an example that raised doubts about Kravetz's main result.

In 1975 Howard Masur [74] showed that Kravetz's result on the curvature of Teichmüller space is false, thereby nullifying Kravetz's proof of the realization problem. To make matters worse, in 1976 Zieschang [129] pointed out that Nielson's original proof for cyclic $\hat{\mathcal{G}}$ is also not correct (see also chapter 5 of [130]).Fortunately, Fenchel's extension to solvable groups $\hat{\mathcal{G}}$ remained valid.

Finally, in 1980 Steve Kerckhoff announced a proof of the general result, but the details were only published in 1983 [60]. In 1981 Eckmann and Müller [23],[24],[25] using methods of homological algebra gave partial answers to the realization problem. Kerckhoff's proof used Thurston's "earthquakes".

It was also known that if one could produce a *non-negative* C^2 strictly Weil-Petersson convex function $\mathcal{F} : \mathcal{T}(M) \to \mathbb{R}$ one could then solve the realization problem (we shall see why in a moment). In 1986, Scott Wolpert [124] proved that appropriate sums of geodesic lengths yield a proper strictly Weil-Petersson convex function on $\mathcal{T}(M)$. His work employs classical results on Kleinian groups due to Fricke and Klein. Recently, the author produced his proof by showing that Dirichlet's energy as a function of the range metric (Wolf's form of Dirichlet's energy) is strictly Weil-Petersson convex.

The interesting aspect of the Dirichlet energy approach is that it employs only methods of partial differential equations and the calculus of variations to solve a long standing problem in two dimensional "discrete" topology.

We shall end this chapter with a proof of the following basic fact:

Theorem 6.4.1 *Suppose there exists a proper, non-negative, C^2 strictly Weil-Petersson convex function $\mathcal{F} : T(M) \to \mathbb{R}^+$. Then the action of any finite subgroup $\hat{\mathcal{G}} \subset \mathcal{D}/\mathcal{D}_0$ on $T(M)$ has a fixed point.*

PROOF: List the elements of $\hat{\mathcal{G}}$ by $[\xi_1], \ldots, [\xi_k]$, with $\xi_j \in \mathcal{D}$. Form a new function $\hat{\mathcal{F}} : T(M) \to \mathbb{R}$ defined by
$$\hat{\mathcal{F}}[g] = \frac{1}{|\hat{\mathcal{G}}|} \sum_i \mathcal{F}\left([\xi_j^* g]\right) \ .$$

Thus $\hat{\mathcal{F}}$ averages $\mathcal{F}$ over $\hat{\mathcal{G}}$. Clearly $\hat{\mathcal{F}}$ is $\hat{\mathcal{G}}$-invariant. Moreover the Weil-Petersson metric comes from the $\mathcal{D}$-equivariant metric $\langle\langle \cdot, \cdot \rangle\rangle$ on $\mathcal{A}$ or $\langle\langle\langle \cdot, \cdot \rangle\rangle\rangle$ on $\mathcal{M}_{-1}$ (see the discussion beginning on page 58). Therefore $\mathcal{D}/\mathcal{D}_0$ acts as a group of *isometries* on $T(M)$, thus preserving Weil-Petersson geodesics. So, $\hat{\mathcal{F}}$ satisfies the hypotheses of lemma 6.3.17 and therefore has a unique critical point, a minimum. This must be a fixed point of $\hat{\mathcal{G}}$, because minima of a $\hat{\mathcal{G}}$-invariant function $\hat{\mathcal{F}}$ automatically come in $\hat{\mathcal{G}}$-orbits. ∎

As a result of this theorem and theorem 6.3.4 that Dirichlet's energy $\tilde{E} : T(M) \to \mathbb{R}^+$ is strictly Weil-Petersson convex, we have the main result of this section:

Theorem 6.4.2 (Nielsen-Realization) *Any finite subgroup $\hat{\mathcal{G}} \subset \mathcal{D}/\mathcal{D}_0$ can be "realized" as a subgroup $\mathcal{G} \subset \mathcal{D}$.*

A Proof of Lichnerowicz' Formula

For completeness, we include here Lichnerowicz' proof for the formula of $DR(g)$. It is taken from [69].

In order to get the correspondence between our notations and definitions and his, note that $\partial_i \sqrt{g} = \sqrt{g} \cdot \Gamma_{ij}^j$ (here and in the following $\partial_i = \frac{\partial}{\partial x^i}$).

PROOF:

$$\partial_i \sqrt{g} = \frac{1}{2\sqrt{g}}\partial_i g = \frac{1}{2\sqrt{g}}\left\{ \det \begin{bmatrix} \partial_i g_{11} & g_{12} & \cdots & g_{1n} \\ \vdots & \vdots & & \vdots \\ \partial_i g_{n1} & g_{n2} & \cdots & g_{nn} \end{bmatrix} + (n-1) \text{ other terms} \right\}$$

$$= \frac{\sqrt{g}}{2}\left\{ \det \begin{bmatrix} g^{11} & \cdots & g^{1n} \\ \vdots & & \vdots \\ g^{n1} & \cdots & g^{nn} \end{bmatrix} \begin{bmatrix} \partial_i g_{11} & g_{12} & \cdots & g_{1n} \\ \vdots & \vdots & & \vdots \\ \partial_i g_{n1} & g_{n2} & \cdots & g_{nn} \end{bmatrix} + (n-1) \text{ other terms} \right\}$$

$$= \frac{\sqrt{g}}{2}\left\{ \det \begin{bmatrix} T_1 & 0 & \cdots & 0 \\ & 1 & & \vdots \\ \vdots & & \ddots & 0 \\ & & & 1 \end{bmatrix} + (n-1) \text{ other terms} \right\} \quad (\text{where } T_1 = \textstyle\sum_j g^{1j}\partial_i g_{j1})$$

$$= \frac{\sqrt{g}}{2}\sum_{j,k} g^{kj}\partial_i g_{jk} = \sqrt{g}\Gamma_{ij}^j$$

where $\Gamma_{ij}^l = \frac{1}{2}g^{kl}(\partial_i g_{kj} + \partial_j g_{ki} - \partial_k g_{ij}) = \frac{1}{2}g^{kl}\Gamma_{ij,k}$.

The covariant derivative ∇_j is defined as follows:

$$\nabla_j T_k^l = \partial_j T_k^l + \Gamma_{js}^l T_k^s - \Gamma_{jk}^s T_s^l \qquad \text{for 1-1 tensors}$$

and analogously for tensors of other type with a similar $+\Gamma T$ term for every upper index and a $-\Gamma T$ term for every lower index. Covariant derivatives of the metric are 0.

As for all other indices, we can raise and lower the index of the covariant derivative: $\nabla^j := g^{jk}\nabla_k$.

Our definition of the divergence agrees with the usual definition

$$(\delta_g h)_j = \nabla^k h_{jk} \ .$$

The Riemann curvature tensor is defined by

$$R^\lambda{}_{\beta\gamma\delta} = \partial_\gamma\Gamma^\alpha_{\beta\delta} - \partial_\delta\Gamma^\alpha_{\beta\gamma} + \Gamma^\alpha_{\rho\gamma}\Gamma^\delta_{\beta\delta} - \Gamma^\alpha_{\rho\delta}\Gamma^\rho_{\beta\gamma} \ .$$

The Ricci curvature tensor is defined by

$$R_{\beta\delta} = R^\alpha{}_{\beta\alpha\delta} \ .$$

It is symmetric.

The scalar curvature is defined by

$$R = R^\alpha_\alpha = g^{\alpha\beta}R_{\alpha\beta} \ .$$

There are different conventions used in the literature. Ours is the same as in Lichnerowicz and in [77].

The Ricci tensor and consequently the scalar curvature differ in sign from Spivak's convention [103]. He uses $R_{\beta\gamma} = R^\alpha{}_{\beta\gamma\alpha}$.

We have already seen that if g is changed in direction h, then

$$Dg^{\alpha\beta} = -h^{\alpha\beta} \ .$$

(For any function of the metric tensor, let $Df := Df(g)h$.)

Now

$$
\begin{aligned}
D\Gamma_{\alpha\beta,\rho} &= \tfrac{1}{2}(\partial_\alpha h_{\beta\rho} + \partial_\beta h_{\alpha\rho} - \partial_\rho h_{\alpha\beta}) \\
&= \tfrac{1}{2}(\nabla_\alpha h_{\beta\rho} + \Gamma^\sigma_{\alpha\beta}h_{\sigma\rho} + \Gamma^\sigma_{\alpha\rho}h_{\beta\sigma} + \nabla_\beta h_{\alpha\rho} + \Gamma^\sigma_{\beta\alpha}h_{\sigma\rho} + \Gamma^\sigma_{\beta\rho}h_{\alpha\sigma} - \\
&\quad -\nabla_\rho h_{\alpha\beta} - \Gamma^\sigma_{\rho\alpha}h_{\sigma\beta} - \Gamma^\sigma_{\rho\beta}h_{\alpha\sigma}) \\
&= \tfrac{1}{2}(\nabla_\alpha h_{\beta\rho} + \nabla_\beta h_{\alpha\rho} - \nabla_\rho h_{\alpha\beta}) + \Gamma^\sigma_{\alpha\beta}h_{\sigma\rho} \ .
\end{aligned}
$$

$$
\begin{aligned}
X^\gamma_{\alpha\beta} := D\Gamma^\gamma_{\alpha\beta} = D(g^{\gamma\rho}\Gamma_{\alpha\beta,\rho}) &= -h^{\gamma\rho}\Gamma_{\alpha\beta,\rho} + g^{\gamma\rho}\left[\tfrac{1}{2}(\nabla_\alpha h_{\beta\rho} + \nabla_\beta h_{\alpha\rho} - \nabla_\rho h_{\alpha\beta}) + \Gamma^\sigma_{\alpha\beta}h_{\sigma\rho}\right] \\
&= -h^\gamma_\rho\Gamma^\rho_{\alpha\beta} + h^\gamma_\sigma\Gamma^\sigma_{\alpha\beta} + \tfrac{1}{2}(\nabla_\alpha h^\gamma_\beta + \nabla_\beta h^\gamma_\alpha - \nabla^\gamma h_{\alpha\beta}) \\
&= \tfrac{1}{2}(\nabla_\alpha h^\gamma_\beta + \nabla_\beta h^\gamma_\alpha - \nabla^\gamma h_{\alpha\beta}) \ .
\end{aligned}
$$

Therefore,

$$DR^\alpha{}_{\beta\gamma\delta} = \partial_\gamma X^\alpha_{\beta\delta} - \partial_\delta X^\alpha_{\beta\gamma} + \Gamma^\alpha_{\rho\gamma} X^\rho_{\beta\delta} + X^\alpha_{\rho\gamma}\Gamma^\rho_{\beta\delta} - \Gamma^\alpha_{\rho\delta} X^\rho_{\beta\gamma} - X^\alpha_{\rho\delta}\Gamma^\rho_{\beta\gamma} \ .$$

Note that X is a tensor although Γ is not. The difference of the Christoffel symbols with respect to two metrics *is* a tensor, and therefore so is the derivative of a Christoffel symbol with respect to the metric. This is obvious from the explicit formula, too. Therefore, the covariant derivative of X is a reasonable concept. Use

$$\partial_\gamma X^\alpha_{\beta\delta} + \Gamma^\alpha_{\gamma\rho} X^\rho_{\beta\delta} - \Gamma^\rho_{\gamma\beta} X^\alpha_{\rho\delta} = \nabla_\gamma X^\alpha_{\beta\delta} + \Gamma^\rho_{\gamma\delta} X^\alpha_{\beta\rho}$$

to get

$$DR^\alpha{}_{\beta\gamma\delta} = \nabla_\gamma X^\alpha_{\beta\delta} - \nabla_\delta X^\alpha_{\beta\gamma} \ .$$

Contracting α with γ and inserting the formula for X gives

$$
\begin{aligned}
DR_{\beta\delta} &= \tfrac{1}{2}\nabla_\alpha(\nabla_\beta h^\alpha_\delta + \nabla_\delta h^\alpha_\beta - \nabla^\alpha h_{\beta\delta}) - \tfrac{1}{2}\nabla_\delta(\nabla_\alpha h^\alpha_\beta + \nabla_\beta h^\alpha_\alpha - \nabla^\alpha h_{\alpha\beta}) \\
&= \tfrac{1}{2}\left(\nabla_\alpha\nabla_\beta h^\alpha_\delta + \nabla_\alpha\nabla_\delta h^\alpha_\beta - \nabla_\alpha\nabla^\alpha h_{\beta\delta} - \nabla_\delta\nabla_\beta(tr\ h)\right) \ .
\end{aligned}
$$

Therefore

$$
\begin{aligned}
DR &= D(g^{\beta\delta} R_{\beta\delta}) = -h^{\beta\delta} R_{\beta\delta} + g^{\beta\delta} DR_{\beta\delta} \\
&= -h^{\beta\delta} R_{\beta\delta} + \tfrac{1}{2}\left(\nabla_\alpha\nabla^\delta h^\alpha_\delta + \nabla_\alpha\nabla^\beta h^\alpha_\beta - \nabla_\alpha\nabla^\alpha(tr\ h) - \nabla^\beta\nabla_\beta(tr\ h)\right) \\
&= -h^{\beta\delta} R_{\beta\delta} + \delta_g\delta_g h - \Delta_g(tr\ h) \ .
\end{aligned}
$$

This formula holds quite generally, but in 2 dimensions, the first term can be considerably simplified.

We claim that in 2 dimensions, $h^{\alpha\beta} R_{\alpha\beta} = \tfrac{1}{2} R \cdot tr_g h$.

The conclusion is done in conformal coordinates. Let

$$g_{11} = g_{22} = \rho, \ \ g_{12} = 0 \ .$$

Then, direct calculations show:

$$
\begin{aligned}
&g^{11} = g^{22} = \rho^{-1}, \ \ g^{12} = 0 \\
&\Gamma^1_{11} = \Gamma^2_{12} = \tfrac{\partial_1\rho}{2\rho}, \ \ \Gamma^2_{22} = \Gamma^1_{12} = \tfrac{\partial_2\rho}{2\rho} \\
&\Gamma^2_{11} = -\tfrac{\partial_2\rho}{2\rho}, \ \ \Gamma^1_{22} = -\tfrac{\partial_1\rho}{2\rho} \\
&R^1{}_{212} = -\tfrac{1}{2}\Delta(\ln\rho) = R_{11} = R_{22}, \ \ R_{12} = 0, \ \ R^1_1 = R^2_2 = -\tfrac{\Delta(\ln\rho)}{2\rho} \\
&R = -\tfrac{\Delta(\ln\rho)}{\rho} \\
&h^{\alpha\beta} R_{\alpha\beta} = h^\alpha_\beta R^\beta_\alpha = h^1_1 R^1_1 + h^2_2 R^2_2 = \tfrac{1}{2} R \, tr_g h
\end{aligned}
\tag{A.1}
$$

B On Harmonic Maps

Introduction

There are now various criteria guaranteeing the existence of harmonic maps between surfaces, and, in general, between Riemannian manifolds. The first major breakthrough was the Eells-Sampson paper [28], mentioned in theorem 3.1.1. As a special case of this result one obtains the existence of a smooth harmonic map from (M, g) to (M, g_0), $g_0 \in \mathcal{M}_{-1}$, and $g \in \mathcal{M}$ an arbitrary metric. The techniques of Eells-Sampson employ the heat flow and therefore involve the study of non-linear parabolic partial differential equations. For a proof of the Eells-Sampson theorem using the Palais-Smale condition via a perturbation argument see Uhlenbeck [119],[120]. For two dimensional surfaces M, N the existence of a smooth harmonic map between M and N in every homotopy class for the case $\pi_2(N) = 0$ was proved by several authors, namely Lemaire [66], Sacks-Uhlenbeck [95] and Brian White [121], and these proofs involve technical analytical results along with the replacement technique of Morrey [78].

Jürgen Jost [57], on the other hand, has given a direct proof of the existence of a harmonic diffeomorphism. We have already encountered the basic compactness idea behind this proof in lemma 3.2.3, namely that diffeomorphisms with bounded Dirichlet energy are equicontinuous (however, the fact that the limit of a minimizing sequence of diffeomorphisms is also a diffeomorphism is non-trivial).

The advantage of the heat flow technique is that it works in arbitrary dimensions and in the case $\partial M \neq \emptyset$ where one assumes Dirichlet boundary data.

We present here our own proofs of existence, uniqueness and smooth dependence. The existence proof follows the spirit of the general approach to Morse theory developed in [111] and we shall say more about this shortly. Our proof of existence can be modified to work, like that of Eells-Sampson, in arbitrary dimensions and in the situation where the domain has non-empty boundary and Dirichlet boundary data are used, when the range is compact and has negative sectional curvature. The case of harmonic maps satisfying Dirichlet boundary conditions was first treated by Hamilton [49] by heat equation methods and for harmonic maps of surfaces by Lemaire [67]. However the most elegant way to treat the general problem of harmonic maps with Dirichlet boundary data is due to Hildebrandt, Kaul, Widman [54],[55],[56] and Giaquinta and Hildebrandt [41]. For an overview of these and other results see the survey paper of Hildebrandt [53].

A harmonic map $S : (M, g) \to (M, g_0)$ is a critical point of Dirichlet's energy for g and g_0 fixed. The Euler-Lagrange equation for harmonic maps is described in section 3.1. How can we find a critical point for E_g? The classical way is to produce an absolute minimum through the direct method of the calculus of variations, the method employed by Jost, Lemaire, Sacks-Uhlenbeck and White.

The idea of Morse theory, as generalized to Hilbert manifolds by Palais and Smale [89], [100], is the method of gradient descent; i.e. follow the flow of a gradient vector field until it leads you to a critical point. This method has, until now, never been made to work in providing a proof of the existence of harmonic mappings. The approach of Eells-Sampson is to follow the trajectories of the heat equation to obtain existence. This method, as beautiful as it is, and as influential as it has been for geometrical problems, requires somewhat more sophisticated methods in non-linear partial differential equations. Our goal is to show that the gradient method (which in reality is the method of ODEs to solve a PDE) works when viewed from the correct prospective. *In doing so, we rely only on the linear theory of elliptic equations, the fundamental existence theorem of local solutions to ordinary differential equations and the Sobolev embedding theorems.*

In the presentation that follows we fix two metrics g and g_0 on M with the scalar curvature of g_0 negative (for Teichmüller theory we need only that $g_0 \in \mathcal{M}_{-1}$, but the proof for general g_0 of negative scalar curvature is the same). Our goal is to first show that a harmonic map $S(g) : (M, g) \to (M, g_0)$ exists, then to show that is is unique and smoothly dependent on g, and finally that $S(g)$ is a diffeomorphism. We begin with:

Existence

For the convenience of the reader we review the Sobolev theorems. Let $L_k^p(M, \mathbb{R}^d)$ denote the space of maps from M into Euclidean space of dimension d which have all partial derivatives (in the sense of distributions) up to k^{th} order in L^p. (Other notations used for the same spaces are $W^{k,p}$ or $H^{k,p}$; in our notation note that $H^k(M) = L_k^2(M, \mathbb{R})$.) In the case $\dim M = 2$, the Sobolev theorems take the following form:

1. If $k - \frac{2}{p} \geq \ell - \frac{2}{p'}$, and $p' < \infty$ there is a continuous inclusion of $L_k^p(M, \mathbb{R}^d)$ into $L_\ell^{p'}(M, \mathbb{R}^d)$. If $k - \frac{2}{p} > \ell - \frac{2}{p}$, and $k > \ell$ this inclusion is completely continuous, or compact.

2. If $k - \frac{2}{p} > \ell$ then there is a completely continuous (= compact) embedding of $L_k^p(M, \mathbb{R}^d)$ into $C^\ell(M, \mathbb{R}^d)$, the space of ℓ times continuously differentiable functions.

For more details the reader should consult [3], [62], [75].

To begin our proof we must introduce the relevant infinite dimensional manifolds of maps. Let $k = 2$, or 3 and let $p > 2$. Assume that M is embedded in $\mathbb{R}^d$. Let $L_k^p(M, M)$ consist of those maps in $L_k^p(M, \mathbb{R}^d)$ which map M to M. (This is well-defined since by the Sobolev theorems these maps will be continuous.) Then standard techniques [90] show that $L_k^p(M, M)$ is a smooth C^∞ manifold and in fact a C^∞ submanifold of $L_k^p(M, \mathbb{R}^d)$. Again by the Sobolev embedding theorem there is a continuous inclusion of $L_2^p(M, M)$ into $C^1(M, M)$, the C^1 self maps on M, and for $k = 3$ a continuous inclusion of $L_3^p(M, M)$ into $C^2(M, M)$, the C^2 self maps of M. In both cases $L_k^p(M, M), k = 2, 3$ are smooth manifolds. It is important to understand their tangent spaces.

The tangent space to $L_k^p(M, M)$ at y consists of the L_k^p vector fields β over u, i.e. those $\beta \in L_k^p(M, \mathbb{R}^d)$ such that $\beta(x) \in T_{u(x)}M$ for all $x \in M$.

We shall now assume that (M, g_0) is isometrically embedded in $\mathbb{R}^d$. For $p \in M$ let

$$\Pi(p) : \mathbb{R}^d \to T_p M$$

be the orthogonal projection. Recall from section 3.1 that $S : (M, g) \to (M, g_0)$ is harmonic iff

$$(\Delta S)(p) = \Pi(S(p))(\Delta_g S)(p) = 0 , \tag{B.1}$$

where Δ_g is the Laplace-Beltrami operator. Define the *non-linear* Laplacian Δ by equation (B.1); S is therefore harmonic iff $\Delta S = 0$. In local conformal coordinates (x^1, x^2) on (M, g),

$$\Delta S = \frac{1}{\sqrt{g}} \frac{D}{\partial x^\ell} \frac{\partial S^\alpha}{\partial x^\ell} , \quad \alpha = 1, \ldots, d . \tag{B.2}$$

For vector fields β over u we can define the *linear Laplacian* Δ (u is assumed fixed) by

$$\Delta \beta = \frac{1}{\sqrt{g}} \frac{D}{\partial x^\ell} \frac{D\beta}{\partial x^\ell} \tag{B.3}$$

where the kth component of the covariant derivative $\frac{D\beta}{\partial x^\ell}$ is given by

$$\left(\frac{D\beta}{\partial x^\ell}\right)^k = \frac{\partial \beta^k}{\partial x^\ell} + \mathring{\Gamma}^k_{i\ell}(u)\beta^i \tag{B.4}$$

where $\mathring{\Gamma}^k_{ij}$ are the Christoffel symbols of g_0.

The spirit of our approach to Morse theory for the calculus of variations taken in [111] was not to stress the gradient nature of the vector field whose trajectories are to lead you to a critical point, but to find the "right" vector field by solving an appropriate linear PDE on the given manifold of maps, in this case $L^p_k(M, M)$.

We are now ready to define our vector field β on $L^p_k(M, M)$ whose trajectories will lead us to a harmonic map. Again fix $u \in L^p_k(M, M)$.

Consider the linear partial differential operator

$$\Delta\beta + \frac{1}{\lambda}\mathcal{R}\left(\beta, \frac{\partial u}{\partial x^\ell}\right)\frac{\partial u}{\partial x^\ell} =: \mathcal{E}_u(\beta) \ . \tag{B.5}$$

Here x^ℓ are conformal coordinates such that $g_{ij} = \lambda\delta_{ij}$. One checks easily that $\beta \mapsto \mathcal{E}_u(\beta)$ maps L^p_k vector field over $u \in L^p_k(M, M)$ to L^p_{k-2} vector fields over u. Moreover $\beta \mapsto \mathcal{E}_u(\beta)$ is a linear self-adjoint second order operator, and therefore by standard elliptic theory the Fredholm alternative holds; i.e. $\beta \mapsto \mathcal{E}_u(\beta)$ is surjective iff it is injective. Another way of saying this is that the operator $\beta \mapsto \mathcal{E}_u(\beta)$ is a linear Fredholm operator of index zero.

Theorem B.1 *For u homotopic to the identity, the map $\beta \mapsto \mathcal{E}_u(\beta)$ is an isomorphism of the L^p_k vector fields over u to the L^p_{k-2} vector fields over u.*

PROOF: By the previous remarks we need only show that $\mathcal{E}_u(\beta) = 0$ implies $\beta = 0$. Suppose $\mathcal{E}_u(\beta) = 0$. Then denoting the $\mathbb{R}^d$ inner product simply by $\langle \cdot, \cdot \rangle$ we have

$$\int_M \langle \mathcal{E}_u(\beta), \beta \rangle \, d\mu_g = 0 \ .$$

Integrating by parts and using the fact that the curvature is negative we obtain the two equations

$$\left\langle \mathcal{R}\left(\beta, \frac{\partial u}{\partial x^\ell}\right)\frac{\partial u}{\partial x^\ell}, \beta \right\rangle = 0 \tag{B.6}$$

and

$$\nabla_{\frac{\partial}{\partial x^\ell}}\beta = 0 \ . \tag{B.7}$$

Since u is C^1 and homotopic to the identity, its degree (which is a homotopy invariant) is 1. By Sard's theorem, u has regular values (i.e. points y such that for every $x \in u^{-1}(y)$, if any, $Du(x)$ is an isomorphism). The degree being 1, these points y do have pre-images. Let us consider some point p_0 for which $Du(p_0)$ is an isomorphism.

Since (M, g_0) has negative curvature equation (B.6) immediately implies that $\beta(p_0) = 0$. Equation (B.7) on the other hand implies that the function $p \mapsto \|\beta(p)\|_{\mathbb{R}^d}^2$ is constant. Thus $\beta \equiv 0$, proving theorem B.1.   ∎

We now define our vector field β on $L_k^p(M, M)$ by

$$\beta(u) := \mathcal{E}_u^{-1} \Delta u \ . \tag{B.8}$$

Since $u \mapsto \mathcal{E}_u$ and $u \mapsto \Delta u$ are smooth, $u \mapsto \beta(u)$ is a smooth (C^∞) vector field on $L_k^p(M, M)$. Moreover it is easy to see that $\beta(u) = 0$ iff u is harmonic. *Thus the zeros of β are precisely the harmonic maps.*

Theorem B.2 *If $S \in L_k^p(M, M)$ is harmonic, the Fréchet derivative $D\beta(S)$ of β at S, is an isomorphism (in fact the identity isomorphism) of $T_S L_k^p(M, M)$, the space of L_k^p vector fields over S.*

PROOF: $\beta(S) = 0$ implies that for $h \in T_S L_k^p(M, M)$ the Fréchet derivative $D\beta(S) : T_S L_k^p(M, M) \to T_S L_k^p(M, M)$ satisfies the equation

$$\Delta D\beta(S)h + \frac{1}{\lambda}\mathcal{R}\left(D\beta(S)h, \frac{\partial S}{\partial x^\ell}\right)\frac{\partial S}{\partial x^\ell} = \Delta h + \frac{1}{\lambda}\mathcal{R}\left(h, \frac{\partial S}{\partial x^\ell}\right)\frac{\partial S}{\partial x^\ell} \ .$$

This last calculation is standard and we leave its verification to the reader. Let $\varphi = D\beta(s)h - h$. Then we have

$$\Delta\varphi + \frac{1}{\lambda}\mathcal{R}\left(\varphi, \frac{\partial S}{\partial x^\ell}\right)\frac{\partial S}{\partial x^\ell} \equiv 0$$

which as in theorem B.1 implies that $\varphi = 0$.   ∎

This last theorem says that each harmonic map is a non-degenerate critical point of Dirichlet's energy E_g in the sense introduced in [111].

Theorem B.3 *The derivative of Dirichlet's energy E_g in the direction β is positive except at a critical point, i.e. $DE_g(u)\beta(u) \geq 0$ and equals zero iff u is harmonic.*

PROOF:

$$DE_g(u)\beta = -\int_M \langle \Delta u, \beta \rangle_{\mathbb{R}^d} d\mu_g$$
$$= -\int_M \left\langle \Delta\beta + \tfrac{1}{\lambda}\mathcal{R}(\beta, \tfrac{\partial u}{\partial x^l})\tfrac{\partial u}{\partial x^l}, \beta \right\rangle d\mu_g \geq 0$$

since $g_0 \in \mathcal{M}_{-1}$. The same argument as was used in theorem B.2 now shows that if $DE_g(u)\beta(u) = 0$ then $\beta(u) = 0$ and hence u is harmonic. ∎

Since β is a smooth vector field, given any initial point $u_0 \in L_k^p(M, M)$ we know by the fundamental existence theorem of ODE's that β has a flow u_t, $t^-(u_0) < t < t^+(u_0)$ with

$$\frac{du_t}{dt} = \beta(u_t) \quad , \quad u_t\big|_{t=0} = u_0 .$$

Our goal is to show that $t^-(u_0) = -\infty$ for all initial conditions $u_0 \in L_3^p(M, M)$, and u_t converges $L_2^p(M, M)$ to a harmonic map as $t \to -\infty$. That this actually happens is suggested by the following two theorems:

Theorem B.4 *Dirichlet's energy $E_g(u_t)$ strictly decreases as t decreases, unless u_0 is harmonic.*

PROOF: $\frac{d}{dt}E_g(u_t) = DE_g(u_t)\frac{du_t}{dt} = DE_g(u_t)\beta(u_t)$, and by theorem B.3 $DE_g(u_t)\beta(u_t) > 0$ unless u_0 is harmonic. ∎

Theorem B.5 $f_t := \Delta u_t$ *satisfies the pointwise exponential equation*

$$\frac{Df_t}{\partial t} = f_t . \tag{B.9}$$

PROOF:

$$\frac{D}{\partial t}\Delta u_t = \Delta\frac{du_t}{dt} + \tfrac{1}{\lambda}\mathcal{R}\left(\frac{du_t}{dt}, \frac{\partial u_t}{\partial x^l}\right)\frac{\partial u_t}{\partial x^l}$$
$$= \Delta\beta + \tfrac{1}{\lambda}\mathcal{R}\left(\beta, \frac{\partial u}{\partial x^l}\right)\frac{\partial u}{\partial x^l}$$
$$= \Delta u_t .$$

∎

Corollary B.6 $\frac{d}{dt}\|\Delta u_t\|_{\mathbb{R}^d}^2 = 2\|\Delta u_t\|_{\mathbb{R}^d}^2.$

Thus we have

$$\|\Delta u_t\|_{\mathbb{R}^d}^2 = e^{2t}\|\Delta u_0\|_{\mathbb{R}^d}^2 .$$

PROOF: Dropping the subscript $\mathbb{R}^d$ from the norm we have

$$\frac{d}{dt}\|\Delta u_t\|^2 = \frac{d}{dt}\|f_t\|^2 = 2\left\langle \frac{Df_t}{\partial t}, f_t \right\rangle = 2\|f_t\|^2 \ .$$

■

Corollary B.7 *As $t \to t^-(u_0)$, $\|\Delta u_t\|^2$ remains pointwise bounded by $\|\Delta u_0\|^2$. If $t \to -\infty$, then $\|\Delta u_t\|^2 \to 0$ pointwise.*   ■

We now continue our proof of the existence of harmonic mappings. Since the theorems we have so far give only pointwise estimates on the *non-linear* Laplacian Δu_t and not on any norm involving the second derivatives of u_t, we cannot yet conclude that $t^-(u_0) = -\infty$. The next step is therefore to work towards such a norm estimate. This will be theorem B.13. Let $e(u)$ denote the energy density of a map $u \in L_k^p(M, M)$. Recall that in conformal coordinates $e(u) = \frac{1}{\lambda}\left\langle \frac{\partial u}{\partial x^i}, \frac{\partial u}{\partial x^i} \right\rangle_{\mathbb{R}^n}$. Then we have the following inequality which will be fundamental to our existence proof. For harmonic maps it is known as the Bochner identity [28] and was used for the existence of harmonic maps by Eells and Sampson.

Theorem B.8
$$\frac{1}{2}\Delta_g e(u) \geq K(g)e(u) + \frac{1}{\lambda^2}\left\langle \frac{D}{\partial x^i}\Delta u, \frac{\partial u}{\partial x^i} \right\rangle$$

where $K(g)$ is the Gauss curvature and $g_{ij} = \lambda \delta_{ij}$.

PROOF: For convenience let us use complex notation

$$2\frac{\partial}{\partial z} = \frac{\partial}{\partial x} - i\frac{\partial}{\partial y} \quad , \quad 2\frac{\partial}{\partial \bar{z}} = \frac{\partial}{\partial x} + i\frac{\partial}{\partial y}$$

$(x, y) = (x^1, x^2)$. Again, using the Einstein summation convention, and again dropping the subscript $\mathbb{R}^d$ from the inner product we see that

$$\begin{aligned}
\Delta_g e(u) &= \frac{4}{\lambda}\frac{\partial}{\partial z}\frac{\partial}{\partial \bar{z}}e(u) \\
&= \frac{4}{\lambda}\frac{\partial}{\partial z}\left\{ -\frac{\lambda_z}{\lambda^2}\left\langle \frac{\partial u}{\partial x^i}, \frac{\partial u}{\partial x^i} \right\rangle + \frac{2}{\lambda}\left\langle \frac{D}{\partial \bar{z}}\frac{\partial u}{\partial x^i}, \frac{\partial u}{\partial x^i} \right\rangle \right\}
\end{aligned}$$

where $2\frac{D}{\partial \bar{z}} = \frac{D}{\partial x} + i\frac{D}{\partial y}$, $2\frac{D}{\partial z} = \frac{D}{\partial x} - i\frac{D}{\partial y}$, and $\lambda_{\bar{z}} = \frac{\partial}{\partial \bar{z}}\lambda$. Continuing we get this equal to

$$\frac{4}{\lambda}\Bigg\{\left(\frac{2\lambda_{\bar{z}}\lambda_z}{\lambda^3} - \frac{\lambda_{\bar{z}z}}{\lambda^2}\right)\left\langle \frac{\partial u}{\partial x^\ell}, \frac{\partial u}{\partial x^\ell}\right\rangle - \frac{2\lambda_{\bar{z}}}{\lambda^2}\left\langle \frac{D}{\partial z}\frac{\partial u}{\partial x^\ell}, \frac{\partial u}{\partial x^\ell}\right\rangle - \frac{2\lambda_z}{\lambda^2}\left\langle \frac{D}{\partial \bar{z}}\frac{\partial u}{\partial x^\ell}, \frac{\partial u}{\partial x^\ell}\right\rangle$$
$$+ \frac{2}{\lambda}\left\langle \frac{D}{\partial z}\frac{\partial u}{\partial x^\ell}, \frac{D}{\partial \bar{z}}\frac{\partial u}{\partial x^\ell}\right\rangle + \frac{2}{\lambda}\left\langle \frac{D}{\partial z}\frac{D}{\partial \bar{z}}\frac{\partial u}{\partial x^\ell}, \frac{\partial u}{\partial x^\ell}\right\rangle\Bigg\} \ .$$

But $\frac{1}{2}\left(\lambda_{\bar{z}}\lambda_z/\lambda^3 - \lambda_{\bar{z}z}/\lambda^2\right) = \frac{1}{4}K(g)$, $K(g)$ the Gauss curvature of the metric g. Using the fact that for a vector field v over u

$$\left(\frac{D}{\partial x}\frac{D}{\partial y} - \frac{D}{\partial y}\frac{D}{\partial y}\right)v = \mathcal{R}\left(\frac{\partial u}{\partial x}, \frac{\partial u}{\partial y}\right)v \ ,$$

we get

$$\begin{aligned}
\frac{1}{8}\Delta_g e(u) &= \frac{1}{4}K(g)e(u) + \frac{\lambda_{\bar{z}}\lambda_z}{2\lambda^4}\left\|\frac{\partial u}{\partial x^\ell}\right\|^2 \\
&\quad - \frac{\lambda_{\bar{z}}}{\lambda^3}\left\langle \frac{D}{\partial z}\frac{\partial u}{\partial x^\ell}, \frac{\partial u}{\partial x^\ell}\right\rangle - \frac{\lambda_z}{\lambda^3}\left\langle \frac{D}{\partial \bar{z}}\frac{\partial u}{\partial x^\ell}, \frac{\partial u}{\partial x^\ell}\right\rangle \\
&\quad + \frac{1}{\lambda^2}\left\langle \frac{D}{\partial z}\frac{\partial u}{\partial x^\ell}, \frac{D}{\partial \bar{z}}\frac{\partial u}{\partial x^\ell}\right\rangle \\
&\quad + \frac{1}{4\lambda^2}\left\langle \mathcal{R}\left(\frac{\partial u}{\partial x^r}, \frac{\partial u}{\partial x^\ell}\right)\frac{\partial u}{\partial x^r}, \frac{\partial u}{\partial x^\ell}\right\rangle + \frac{1}{4\lambda^2}\left\langle \frac{D}{\partial x^\ell}\left(\frac{D}{\partial x^r}\frac{\partial u}{\partial x^r}\right), \frac{\partial u}{\partial x^\ell}\right\rangle \ .
\end{aligned} \tag{B.10}$$

Now if $u : M \to M$, we can take the covariant derivative ∇Du of $Du : T_xM \to T_{u(x)}M$. So $Du \in \mathrm{Hom}(TM, u^*TM)$ and $\nabla DU \in \mathrm{Hom}(TM, \mathrm{Hom}(TM, u^*TM)) \cong \mathrm{Hom}(TM \otimes TM, u^*TM)$. ∇Du is defined by

$$\nabla Du = A^\gamma_{\alpha\beta}dx^\alpha \otimes dx^\beta \otimes \frac{\partial}{\partial y^\gamma}\Big|_u \ ,$$

where

$$A^\gamma_{\alpha\beta} = \left(\frac{\partial^2 u^\gamma}{\partial x^\alpha \partial x^\beta} - \frac{\partial u^\gamma}{\partial x^\rho}\Gamma^\rho_{\alpha\beta} + \frac{\partial u^i}{\partial x^\beta}\frac{\partial u^j}{\partial x^\alpha}\mathring{\Gamma}^\gamma_{ji}\right)$$

and $\Gamma^\rho_{\alpha\beta}$ and $\mathring{\Gamma}^\gamma_{ji}$ are the Christoffel symbols of g and g_0 respectively. Thus equations (B.10) can be rewritten as

$$\begin{aligned}
\frac{1}{8}\Delta_g e(u) &= \frac{1}{4}K(g)e(u) + \frac{1}{4\lambda^2}\left\langle \mathcal{R}\left(\frac{\partial u}{\partial x^r}, \frac{\partial u}{\partial x^\ell}\right)\frac{\partial u}{\partial x^r}, \frac{\partial u}{\partial x^\ell}\right\rangle \\
&\quad + \frac{1}{4}|\nabla Du|^2 + \frac{1}{4\lambda^2}\left\langle \frac{D}{\partial x^\ell}\frac{D}{\partial x^r}\frac{\partial u}{\partial x^r}, \frac{\partial u}{\partial x^\ell}\right\rangle
\end{aligned}$$

where $|\nabla Du|^2 = (g_0)_{ij}g^{\alpha\rho}g^{\mu\beta}A^i_{\alpha\beta}A^j_{\rho\mu}$.

Using the fact that g_0 has negative curvature we see that

$$\frac{1}{2}\Delta_g e(u) \geq K(g)e(u) + \frac{1}{\lambda^2}\left\langle \frac{D}{\partial x^\ell}\frac{D}{\partial x^r}\frac{\partial u}{\partial x^r}, \frac{\partial u}{\partial x^\ell}\right\rangle \ . \tag{B.11}$$

This ends the proof. ■

The concept of a covariant derivative employed here is a bit more subtle than the usual textbook definition, which is good enough everywhere else in this book. It is, however, not really necessary to understand the geometrical meaning of ∇Du here. The only thing we use is that the term called $|\nabla Du|^2$ in the above formula is non-negative.

Lemma B.9 *Let $\varphi : M \to \mathbb{R}$ be a C^∞ test function. Let u_0 be an initial value for the gradient flow $\frac{du_t}{dt} = \beta(u_t)$, $t \in]t^-(u_0), 0]$. Let $c_0 = E_g(u_0)$ and $c_1 = \sup_{(t,p)} \|\Delta u_t(p)\| = \sup_p \|\Delta u_0(p)\|$. We then have the following inequality:*

$$-\frac{1}{2}\int_M \langle \nabla_g e(u_t), \nabla_g\varphi \rangle \, d\mu_g \geq -c_2 \int_M e(u_t)\,|\varphi|d\mu_g - \frac{3c_1}{2}\int_M |\varphi|d\mu_g - c_1(2c_0)^{1/2}\|\nabla_g\varphi\|_{L^2}$$

where $\|\nabla_g\varphi\|_{L^2}$ is the L^2-norm of the gradient of φ.

PROOF: Multiplying equation (B.11) through by φ and integrating by parts we get

$$
\begin{aligned}
-\tfrac{1}{2}\int_M \langle \nabla_g e(u_t), \nabla_g\varphi \rangle \, d\mu_g \geq\ & \int_M K(g)e(u_t)\varphi d\mu_g - \int \|\Delta u_t\|^2 \varphi d\mu_g \\
& + \int_M \left\langle \Delta u_t, \tfrac{\partial u}{\partial x^\ell} \right\rangle \left(\tfrac{\lambda_{x^\ell}}{\lambda^2}\right)\varphi d\mu_g \\
& - \int_M \left\langle \Delta u_t, \tfrac{\partial u}{\partial x^\ell} \right\rangle \tfrac{1}{\lambda}\left(\tfrac{\partial\varphi}{\partial x^\ell}\right) d\mu_g \\
\geq\ & -\int |K(g)|\,e(u_t)\,|\varphi|d\mu_g - c_1 \int_M |\varphi|d\mu_g \\
& - \int \tfrac{1}{2}\left(\|\Delta u_t\|^2 \cdot \left|\tfrac{\lambda^2}{\lambda_{x^\ell}}\right| + \left\|\tfrac{\partial u}{\partial x^\ell}\right\|^2 \left|\tfrac{\lambda_{x^\ell}}{\lambda^2}\right|\right)\cdot\left|\tfrac{\lambda_{x^\ell}}{\lambda^2}\right| |\varphi|d\mu_g \\
& - c_1(2c_0)^{1/2}\|\nabla_g\varphi\|_{L^2} \\
\geq\ & -\int_M |K(g)|\,e(u_t)\,|\varphi|d\mu_g - \tfrac{3}{2}c_1 \int_M |\varphi|d\mu_g \\
& - \tfrac{1}{2}\int e(u_t)\left\{\tfrac{\lambda_x^2+\lambda_y^2}{\lambda^3}\right\}|\varphi|d\mu_g - c_1(2c_0)^{1/2}\|\nabla_g\varphi\|_{L^2} \ .
\end{aligned}
$$

Covering (M, g) with a finite number of coordinate charts U_i, W_i, $U_i \supset \overline{W_i}$ such that $\bigcup W_i = M$ we see that we can bound $(\lambda_x^2 + \lambda_y^2)/\lambda^3$ by some positive constant on each $\overline{W_i}$. Therefore we can bound

$$\frac{1}{2}\int_M e(u_t)\frac{\lambda_x^2 + \lambda_y^2}{\lambda^3}|\varphi|\, d\mu_g$$

by $c \int_M e(u_t)|\varphi|\, d\mu_g$ where c is some positive constant depending on (M, g). Combining this with the expression $\int_M |K(g)|\,e(u_t)\,|\varphi|\, d\mu_g$ we can bound the sum of the first and the third terms immediately above by $-c_2 \int_M e(u_t) |\varphi| \, d\mu_g$. Therefore the above inequality can be written as

$$-\frac{1}{2} \int_M \langle \nabla_g e(u_t), \nabla_g \varphi \rangle \, d\mu_g \geq -c_2 \int_M e(u_t) |\varphi| \, d\mu_g - \frac{3}{2} c_1 \int_M |\varphi| d\mu_g - c_1 (2c_0)^{1/2} ||\nabla_g \varphi||_{L^2} \; .$$

∎

This permits us to prove:

Theorem B.10 *Let u_0, c_2, c_1 and c_0 be as in lemma B.9. Then for $t \in]t^-(u_0), 0]$ and for any $\tilde{q}$, $1 < \tilde{q} < 2$ we have*

$$||\nabla_g e(u_t)||_{L^{\tilde{q}}} \leq c_3 \tag{B.12}$$

where the constant c_3 depends only on u_0, the manifold (M, g), the constants c_0, c_1, c_2 and the Sobolev constant $c_{\tilde{q}}$ coming from the embedding of $L_1^{\tilde{p}}$ into L^∞, $\frac{1}{\tilde{p}} + \frac{1}{\tilde{q}} = 1$; i.e.

$$||\varphi||_\infty \leq c_{\tilde{p}} ||\varphi||_{L_1^{\tilde{p}}} \; ,$$

$||\varphi||_\infty$ *the L^∞ norm of φ.*

Corollary B.11 *For any $\tilde{q}$, $1 \leq \tilde{q} < 2$ the $L_1^{\tilde{q}}$ norm of $e(u_t)$ is bounded.*

PROOF OF COROLLARY: An easy exercise in the use of the Sobolev embedding theorems shows that $\int_M e(u_t) d\mu_g < \infty$ and (B.12) imply that $e(u_t)$ is bounded in $L_1^{\tilde{q}}$.

PROOF OF B.10: By lemma B.9 we have

$$\begin{aligned}
\int_M \langle \nabla_g e(u_t), \nabla_g \varphi \rangle \, d\mu_g \; &\leq \; 2c_2 c_0 ||\varphi||_\infty + 3c_1 ||\varphi||_\infty \, \text{area}(M) \\
&\quad + 2c_1 (2c_0)^{1/2} ||\nabla_g \varphi||_{L^{\tilde{p}}} \cdot \text{area}(M)^{2\tilde{p}/(\tilde{p}-2)} \\
&\leq \; \left(2c_2 c_{\tilde{p}} c_0 + 3c_1 c_{\tilde{p}} \text{area}(M) + 2c_1 (2c_0)^{1/2} \text{area}(M)^{2\tilde{p}/(\tilde{p}-2)} \right) ||\varphi||_{L_1^{\tilde{p}}} \\
&\leq \; c_3 ||\varphi||_{L_1^{\tilde{p}}} \; .
\end{aligned}$$

If φ is chosen so that $||\varphi||_{L_1^{\tilde{p}}} \leq 1$ we see that

$$\int_M \langle \nabla_g e(u_t), \nabla_g \varphi \rangle \, d\mu_g \leq c_3$$

for all such φ. This clearly implies that

$$||\nabla_g e(u_t)||_{L^{\tilde{q}}} \leq c_3 \; .$$

∎

Corollary B.12 *For any p, $1 \leq p < \infty$ the first derivatives $\frac{\partial u_t}{\partial x^\ell}$ are bounded uniformly in L^p for $t \in]t^-(u_0), 0]$ in any fixed conformal coordinate chart.*

PROOF: By the Sobolev embedding theorems, given any p there is a $\tilde{q}$, $1 < \tilde{q} < 2$ such that there is a continuous inclusion of $L_1^{\tilde{q}}(M, \mathbb{R})$ into $L^p(M, \mathbb{R})$. Thus $e(u_t)$ are bounded in L^p for any $p \geq 1$ and hence $\frac{\partial u_t}{\partial x^\ell}$ are bounded in L^p. ∎

We have now found the analytical key for producing a harmonic map, namely

Theorem B.13 *For any p, $1 \leq p < \infty$ and $t \in]t^-(u_0), 0]$ the flow u_t remains bounded in the L_2^p norm.*

PROOF: Recall that $u \in L_2^p(M, M)$ is harmonic if in a conformal coordinate system

$$(\Delta u)^k = \Delta_g u^k + \Gamma_{ij}^k(u)\frac{\partial u^i}{\partial x^\alpha}\frac{\partial u^j}{\partial x^\alpha} = 0 \tag{B.13}$$

where $\Delta_g u^k$ is the Laplace-Beltrami operator of the kth component of $u : M \to \mathbb{R}^d$ and $(\Delta u)^k$ is the kth conmponent of the non-linear Laplacian. Along the flow u_t we have

$$- \Delta_g u_t^k + u_t^k = u_t^k + \Gamma_{ij}^k(u)\frac{\partial u^i}{\partial x^\alpha}\frac{\partial u^j}{\partial x^\alpha} - (\Delta u_t)^k \ . \tag{B.14}$$

Now $-\Delta_g + I$, I the identity, is a strictly positive operator. But by corollaries B.7 and B.11 the right hand side of (B.14) is bounded in L^p for any p. Therefore by standard elliptic estimates u_t is bounded in L_2^p for any p. ∎

Our next goal is to show that $\sup \left\{ \|\beta(u_t)(p)\| \ \middle| \ p \in M, \ t \in]t^-(u_0), 0] \right\} < \infty$; i.e. the vector field β is bounded along the trajectories. We begin with

Lemma B.14 *Let $u \in L_2^p(M, M)$. Then for all tangent vectors $X \in T_u L_2^p(M, M)$, there exists a constant $c > 0$ such that*

$$\int_M \|X\|^2 \, d\mu_g \leq c \sum_\ell \left\{ \int_M \left\| \frac{DX}{\partial x^\ell} \right\|^2 \, dx \, dy - \int_M \left\langle \mathcal{R}\left(X, \frac{\partial u}{\partial x^\ell}\right)\frac{\partial u}{\partial x^\ell}, X \right\rangle \, dx \, dy \right\}$$

where X (a priori a vector field over u) is considered as a map from M into the ambient space $\mathbb{R}^d$.

PROOF: We argue by contradiction. If not, there is a sequence X_m, $\int\limits_M \|X_m\|^2 \, d\mu_g = 1$ with

$$1 \geq m \sum_\ell \int\limits_M \left\{ \left\| \frac{DX_m}{\partial x^\ell} \right\|^2 - \left\langle \mathcal{R}\left(X_m, \frac{\partial u}{\partial x^\ell}\right) \frac{\partial u}{\partial x^\ell}, X_m \right\rangle \right\} dx \, dy \ .$$

Thus $\sum_\ell \int\limits_M \left\| \frac{DX_m}{\partial x^\ell} \right\|^2 dx \, dy \to 0$, $\sum_\ell \left\langle \mathcal{R}\left(X_m, \frac{\partial u}{\partial x^\ell}\right) \frac{\partial u}{\partial x^\ell}, X_m \right\rangle \to 0$ and $\int\limits_M \|X_n\|^2 \, d\mu_g = 1$.

Using the fact that

$$\frac{DX}{\partial x^\ell} = \frac{\partial X}{\partial x^\ell} - D\Pi(u)\left(\frac{\partial u}{\partial x^\ell}\right) X \tag{B.15}$$

where $\Pi(p) : \mathbb{R}^d \to T_p M$ is the orthogonal projection, we see that X_m is bounded in the L_1^2 norm as a map from M into $\mathbb{R}^d$. By standard functional analysis X_m has a subsequence, say X_m again, which converges weakly to some X in $L_1^2(M, \mathbb{R}^d)$. Since by the Sobolev embedding theorem, the inclusion of $L_1^2(M, \mathbb{R}^d)$ into $L_0^2(M, \mathbb{R}^d) := L^2(M, \mathbb{R}^d)$ is compact, we may assume that X_m converges in the L^2-norm to X.

However, this implies that

$$\sum_\ell \left\langle \mathcal{R}\left(X, \frac{\partial u}{\partial x^\ell}\right) \frac{\partial u}{\partial x^\ell}, X \right\rangle \equiv 0$$

at almost all points $p \in M$. This implies that $X = 0$ contradicting the fact that $\|X\|_{L^2} = 1$. ∎

This lemma can be strengthened so that the constant C can be selected to be fixed along a trajectory u_t, $t \in]t^-(u_0), 0]$. Thus we have:

Lemma B.15 *Let u_t be a trajectory of β, $t \in]t^-(u_0), 0]$. Then there is a constant $C > 0$ such that for all tangent vectors $X_t \in T_{u_t} L_3^p(M, M)$*

$$\int\limits_M \|X_t\|^2 \, d\mu_g \leq C \sum_\ell \left\{ \int\limits_M \left\| \frac{DX_t}{\partial x^\ell} \right\|^2 dx \, dy - \int \left\langle \mathcal{R}\left(X_t, \frac{\partial u_t}{\partial x^\ell}\right) \frac{\partial u_t}{\partial x^\ell}, X_t \right\rangle dx \, dy \right\} \ .$$

PROOF: We again argue by contradiction picking a subsequence X_{t_m} such that $\|X_{t_m}\|_{L^2}^2 = 1$ and

$$1 \geq m \sum_\ell \left\{ \int\limits_M \left\| \frac{DX_{t_m}}{\partial x^\ell} \right\|^2 dx \, dy - \int \left\langle \mathcal{R}\left(X_{t_m}, \frac{\partial u_{t_m}}{\partial x^\ell}\right) \frac{\partial u_{t_m}}{\partial x^\ell}, X_{t_m} \right\rangle dx \, dy \right\} \ .$$

By theorem B.13 u_{t_m} is bounded in the L_2^p norm for any p. By the Sobolev embedding theorem we can extract a subsequence, again called u_{t_m} which converges C^1 to $u \in L_2^p$, in particular u_{t_m} are bounded C^1. This allows us to conclude, as in B.15 that X_{t_m} is bounded in $L_1^2(M, \mathbb{R}^d)$. Thus we may choose a subsequence X_{t_m} which converges weakly to X in L_1^2 and strongly in L^2. As before we conclude that $X = 0$ and $\|X\|_{L^2}^2 = 1$, a contradiction. ∎

As a direct consequence of B.15 we have

Lemma B.16 *The vector field β is bounded in L^2 along any trajectory u_t, for $t \in]t^-(u_0), 0]$, with the bound independent of t.*

PROOF: Recall that the derivative $DE_g(u_t)$ of Dirichlet's energy is given by

$$
\begin{aligned}
DE_g(u_t)\beta &= \sum_\ell \int_M \left\langle \frac{Du_t}{\partial x^\ell}, \frac{D\beta}{\partial x^\ell} \right\rangle dx \, dy \\
&= -\int_M \langle \Delta u_t, \beta \rangle \, d\mu_g \\
&= -\int_M \langle \Delta\beta, \beta \rangle \, d\mu_g - \sum_\ell \int_M \left\langle \mathcal{R}(\beta, \frac{\partial u_t}{\partial x^\ell})\frac{\partial u_t}{\partial x^\ell}, \beta \right\rangle dx \, dy \\
&= \sum_\ell \left\{ \int_M \left\langle \frac{D\beta}{\partial x^\ell}, \frac{D\beta}{\partial x^\ell} \right\rangle dx \, dy - \int_M \left\langle \mathcal{R}\left(\beta, \frac{\partial u_t}{\partial x^\ell}\right)\frac{\partial u_t}{\partial x^\ell}, \beta \right\rangle dx \, dy \right\}
\end{aligned}
$$

where we write $\beta := \beta(u_t)$.

But

$$
\begin{aligned}
\sum_\ell \int_M \left\langle \frac{\partial u_t}{\partial x^\ell}, \frac{D\beta}{\partial x^\ell} \right\rangle dx \, dy &\leq \sum_\ell \int_M \frac{1}{2}\left(\left\|\frac{\partial u_t}{\partial x^\ell}\right\|^2 + \left\|\frac{D\beta}{\partial x^\ell}\right\|^2 \right) dx \, dy \\
&= E_g(u_t) + \frac{1}{2}\sum_\ell \int_M \left\|\frac{D\beta}{\partial x^\ell}\right\|^2 dx \, dy \ .
\end{aligned}
$$

Thus

$$
\frac{1}{2}\sum_\ell \int_M \left\|\frac{D\beta}{\partial x^\ell}\right\|^2 dx \, dy - \sum_\ell \int_M \left\langle \mathcal{R}\left(\beta, \frac{\partial u_t}{\partial x^\ell}\right)\frac{\partial u_t}{\partial x^\ell}, \beta \right\rangle dx \, dy \leq E_g(u_t) \leq E_g(u_0)
$$

for all $t \in]t^-(u_0), 0]$. Applying the inequality of lemma B.15 completes the proof. ∎

Theorem B.17 *Along a trajectory u_t, $t \in]t^-(u_0), 0]$ the vector field β is bounded in supremum norm by $\tilde{C}\|\beta\|_{L^2}$, where $\tilde{C}$ is a positive constant depending only on u_0 and (M, g).*

PROOF: β is a vector field in $L_3^p(M, M)$, therefore it is C^2. Fix t and consider $||\beta||^2 :=$ $||\beta(u_t)(p)||^2$ as a real valued C^2 function of $p \in M$. Taking the Laplace-Beltrami operator we obtain

$$\frac{1}{2}\Delta_g ||\beta||^2 = \sum_\ell \frac{1}{\lambda} \left\| \frac{D\beta}{\partial x^\ell} \right\|^2 + \langle \Delta\beta, \beta \rangle \geq \langle \Delta u_t, \beta \rangle - \sum_\ell \frac{1}{\lambda} \left\langle \mathcal{R}\left(\beta, \frac{\partial u_t}{\partial x^\ell}\right) \frac{\partial u_t}{\partial x^\ell}, \beta \right\rangle$$

where again $g_{ij} = \lambda \delta_{ij}$.

Since g_0 has negative curvature we get that

$$\Delta_g ||\beta||^2 \geq 2 \langle \Delta u_t, \beta \rangle \ .$$

But $\sup_{(p,t)} ||\Delta u_t(p)|| \leq c_1 := \sup_p ||\Delta u_0(p)||$ by corollary B.6. Applying the Schwarz inequality we see that

$$\Delta_g ||\beta||^2 \geq -c_1 - ||\beta||^2 \ .$$

A standard fact from linear PDE states that if a C^2 function $\varphi : M \to \mathbb{R}$ satisfies $\Delta_g \varphi \geq -C\varphi$ then the sup norm of φ is bounded by a constant $\tilde{C}$ times the L^2-norm where $\tilde{C}$ depends only on M (see for example the paper by F. Tomi [109]). Putting $\varphi = ||\beta||^2 + c_1$ we conclude that

$$\sup_p ||\beta(u_t)(p)|| \leq \tilde{C} ||\beta||_{L^2} \ ,$$

where $\tilde{C}$ depends only on M and c_1 and is therefore independent of t.

Now using lemma B.16 we can immediately conclude

Lemma B.18 *There is a constant c_4 independent of t such that*

$$\sup_p ||\beta(u_t)(p)|| \leq c_4 \ , \quad t \in]t^-(u_0), 0] \ .$$

∎

Our next to final step is the following

Lemma B.19 *Let $p = 2m$, $m \geq 1$. Then Δu_t is bounded in the L_1^p norm with the bound independent of $t \in]t^-(u_0), 0]$.*

PROOF: Since

$$\Pi(u_t)\frac{\partial}{\partial x^\ell}\Delta_g u_t = \frac{D}{\partial x^\ell}(\Delta u_t) - \left(D\Pi(u_t)\frac{\partial u_t}{\partial x^\ell}\right)\Delta_g u_t$$

and both $\frac{\partial u_t}{\partial x^\ell}$ and Δu_t are bounded in supremum norm it suffices to prove that the intrinsic quantity $\lambda^{-1/2}\frac{D}{\partial x^\ell}\Delta u_t$ is bounded in L^p. Let $f_t = \Delta u_t$ and

$$\varphi(t) := \int_M \frac{1}{\lambda^m}\left\|\frac{D}{\partial x^\ell}f_t\right\|^{2m} d\mu_g = \int_M \frac{1}{\lambda^m}\left\langle\frac{D}{\partial x^\ell}f_t, \frac{D}{\partial x^\ell}f_t\right\rangle^m d\mu_g \ .$$

Therefore differentiating with respect to t we obtain

$$\frac{d\varphi}{dt} = m\varphi(t) + m\int_M \frac{1}{\lambda^m}\left\langle\frac{Df_t}{\partial x^\ell}, \frac{Df_t}{\partial x^\ell}\right\rangle^{m-1}\left\langle\mathcal{R}\left(\beta, \frac{\partial u}{\partial x^\ell}\right)f_t, \frac{D}{\partial x^\ell}f_t\right\rangle d\mu_g \ .$$

Since both f_t and β are bounded in sup norm by some universal constant, as is the intrinsic quantity $\lambda^{-1/2}\partial u/\partial x^\ell$, we may conclude that

$$\left|\frac{m}{\lambda^{m-1/2}}\left\langle\mathcal{R}\left(\beta, \frac{1}{\sqrt{\lambda}}\frac{\partial u}{\partial x^\ell}\right)f_t, \frac{D}{\partial x^\ell}f_t\right\rangle\right| \geq -c_5\frac{1}{\lambda^{m-1/2}}\left\|\frac{Df_t}{\partial x^\ell}\right\| \ .$$

Thus we see that

$$\frac{d\varphi}{dt} \geq m\varphi(t) - c_5\int_M \frac{1}{\lambda^{m-1/2}}\left\|\frac{Df_t}{\partial x^\ell}\right\|^{2m-1} d\mu_g \ .$$

By the Schwarz inequality

$$\int_M \frac{1}{\lambda^{m-1/2}}\left\|\frac{Df_t}{\partial x^\ell}\right\|^{2m-1} d\mu_g \leq C\left\{\int_M \frac{1}{\lambda^m}\left\|\frac{Df_t}{\partial x^\ell}\right\|^{2m}\right\}^{(2m-1)/2m} = C\varphi(t)^\alpha \ , \quad \alpha = \frac{2m-1}{2m} < 1 \ .$$

Consequently we arrive at the inequality

$$\frac{d\varphi}{dt} \geq m\varphi(t) - c_6\varphi(t)^\alpha \ .$$

Let $\psi^\gamma := \varphi$ where $\gamma = \frac{1}{1-\alpha}$. Then

$$\gamma\psi^{\gamma-1}\frac{d\psi}{dt} = \frac{d\varphi}{dt} \geq m\varphi - c_6\varphi^\alpha = m\psi^\gamma - c_6\psi^{\alpha\gamma} \ .$$

Thus

$$\gamma \cdot \frac{d\psi}{dt} \geq m\psi - c_6\psi^{\alpha\gamma-\gamma+1} \ .$$

But $\alpha\gamma - \gamma + 1 = 0$ and so we arrive at the differential inequality

$$\frac{d\psi}{dt} \geq \left(\frac{m}{\gamma}\right)\psi - c_6 \ , \quad \psi \geq 0 \ . \tag{B.16}$$

We only have to show that ψ is bounded for all negative time; then φ will also be bounded, finishing the proof. We claim that the bound for ψ is $\max(c_6\gamma/m, \psi(0))$. This bound is independent of $t \in]t^-(u_0), 0]$. If $\psi(0) > c_6\gamma/m$, then according to (B.16) ψ will decrease as t decreases, either forever (which finishes the proof) or until $\psi(t_0) \leq c_6\gamma/m$ for some $t_0 \leq 0$. This is already the situation in the other case $\psi(0) \leq c_6\gamma/m$. But then, no $t < t_0$ can satisfy $\psi(t) > c_6\gamma/m$, because in this case an interval $[t_{-1}, t_0']$ would exist on which $\psi \geq c_6\gamma/m$ but $\psi(t_0') = c_6\gamma/m$, $\psi(t_{-1}) > c_6\gamma/m$. Integrating (B.16) over this interval leads to a contradiction. ∎

We already know (theorem B.13) that u_t, $t \in]t^-(u_0), 0]$ is bounded in L_2^p for any p with the bound independent of t. Combining formula (B.13) and lemma B.19 we can immediately conclude that u_t is bounded in L_3^p for any $p = 2m$ again with the bound independent of negative time. We state this formally as

Theorem B.20 *For any $p = 2m$, the flow u_t of β is bounded in L_3^p, with the bound independent of negative time.*

Now let us consider the L_3^p norm of the vector field β. We have

Theorem B.21 *The vector field β is bounded in the L_3^p norm along a trajectory u_t, $t \in]t^-(u_0), 0]$ for any $p = 2m$, with the bound independent of t.*

PROOF: The elliptic equation for β is in conformal coordinates $g_{ij} = \lambda\delta_{ij}$

$$\Delta\beta + \frac{1}{\lambda}\mathcal{R}\left(\beta, \frac{\partial u_t}{\partial x^l}\right)\frac{\partial u_t}{\partial x^l} = \Delta u_t \ , \tag{B.17}$$

where again $\beta = \beta(u_t)$. If $\Pi(p) : \mathbb{R}^n \to T_pM$ is the g_0-orthogonal projection then

$$\Delta\beta = \frac{1}{\lambda}\Pi\frac{\partial}{\partial x^l}\Pi\frac{\partial}{\partial x^l}\beta \ . \tag{B.18}$$

Using the fact that $\Pi(u(p))\beta(u)(p) = \beta(u)(p)$, (B.18) can be written as

$$\begin{aligned}
\lambda\Delta\beta \ = \ & \frac{\partial^2\beta}{\partial x^{l2}} - D^2\Pi(u_t)\left(\frac{\partial u_t}{\partial x^l}\right)\left(\Pi\frac{\partial u_t}{\partial x^l}\right)\beta \\
& - D\Pi(u_t)\left[D\Pi(u_t)\left(\frac{\partial u_t}{\partial x^l}\right)\right]\left(\frac{\partial u_t}{\partial x^l}\right)\beta \\
& - D\Pi(u_t)\left[\Pi\frac{\partial^2 u_t}{\partial x^{l2}}\right]\beta - D\Pi(u_t)\left(\frac{\partial u_t}{\partial x^l}\right)\frac{\partial\beta}{\partial x^l} \\
& - D\Pi(u_t)\left(\frac{\partial u_t}{\partial x^l}\right)\left[\frac{\partial\beta}{\partial x^l} - D\Pi(u_t)\left(\frac{\partial u_t}{\partial x^l}\right)\beta\right] \ .
\end{aligned}$$

We know that $\|\beta\|^2$ is bounded in sup norm and from the proof of lemma B.16 we see that the intrinsic quantity $\frac{1}{\sqrt{\lambda}}\left\|\frac{D\beta}{\partial x^l}\right\|$ is bounded in L^2. From equation (B.15) and the fact that $\frac{1}{\sqrt{\lambda}}\left\|\frac{\partial u_t}{\partial x^l}\right\|$ is bounded in sup norm we can conclude that $\frac{1}{\sqrt{\lambda}}\left\|\frac{\partial \beta}{\partial x^l}\right\|$ is bounded in L^2. Therefore we may rewrite equation (B.18) as

$$-\Delta_g \beta + \beta = \Omega \qquad\qquad (B.19)$$

where $\Delta_g \beta$ is the Laplace-Beltrami operator on the vector $\beta = (\beta^1, \ldots, \beta^n)$ and Ω is bounded in L^2. Equation (B.19) is essentially an elliptic equation in each component β^j with the right side Ω^j bounded in L^2. Therefore, since $-\Delta_g + I$ is a strictly positive invertible operator, we may conclude from standard elliptic estimates that β is bounded in the L_2^2 norm. From the Sobolev embedding theorems it follows that $\frac{1}{\sqrt{\lambda}}\frac{D\beta}{\partial x^l}$ is bounded in L_1^2. Since Δu_t is in L_1^2 it follows that the Ω in (B.19) is in L_1^2. Thus $\beta \in L_3^2$. Again using the Sobolev inequalities we conclude that $\frac{1}{\sqrt{\lambda}}\frac{D\beta}{\partial x^l} \in L_1^p$ for any p. Since $\Delta u_t \in L_1^p$ for any $p = 2m$, looking once again at (B.19) we see that $\beta \in L_3^p$, $p = 2m$. ∎

Theorem B.22 *The flow $u_t \in L_3^p$, $p = 2m > 2$ is defined for all time.*

PROOF: From theorem B.21 β is bounded in the L_3^p norm and a standard result [64] from ODE is that if a vector field on a complete manifold ($L_3^p(M, M)$ is complete, since M is compact) is bounded along a trajectory u_t, $t \in]t^-(u_0), 0]$, then necessarily $t^-(u_0) = -\infty$.

We have now come to the result we have worked towards:

Theorem B.23 *For any metric g on M and any metric g_0 on M with negative scalar curvature there exists a C^∞ smooth harmonic map $S : (M, g) \to (M, g_0)$ homotopic to the identity.*

PROOF: By theorem B.22 the flow u_t goes for all negative time. From the equation

$$\|\Delta u_t\|^2 = e^{2t}\|\Delta u_0\|^2$$

we see that $\|\Delta u_t\|^2 \to 0$ as $t \to -\infty$. We also know that u_t is bounded in L_3^p for any $p = 2m$. If $p > 2$ then the inclusion of L_3^p into C^2 is compact. If $t_m \to -\infty$ we can therefore extract a subsequence say t_m such that u_{t_m} converges C^2 to some C^2 map S. Clearly S must be harmonic. Write the equation for harmonic maps in conformal coordinates as

$$\Delta_g S - \frac{1}{\lambda} D\Pi(S)\left(\frac{\partial S}{\partial x^l}\right)\frac{\partial S}{\partial x^l} = 0$$

or

$$-\Delta_g S + S = \Omega'$$

where Ω' is C^1. Again, repeated application of elliptic regularity for $-\Delta_g + I$ yields that $S \in C^\infty$. ∎

Uniqueness of Harmonic Maps

Consider Dirichlet's energy E_g as a real valued smooth map on $L_3^p(M, M)$. Then the fact that $g_0 \in \mathcal{M}_{-1}$ (or simply that the scalar curvature of g_0 is negative) implies that every critical point of E_g is a (non-degenerate) minimum. To show uniqueness one could proceed as follows:

1. Show that $E_g : L_3^p(M, M) \to \mathbb{R}$ satisfies the axioms for a Morse theory as described in [111]

2. Since $L_3^p(M, M)$ is connected one can then use a mountain pass argument to show that the existence of two non-degenerate minima implies the existence of at least one other critical point which is not a minimum, establishing a contradiction.

This procedure would involve checking that the Morse theory of [111] holds in the case at hand. Although this is true, there is, fortunately, a much shorter and simpler way of establishing uniqueness.

Suppose we have two harmonic maps S_0 and S_1 mapping (M, g) to (M, g_0). Since S_0 and S_1 are homotopic there is a smooth homotopy $F : M \times I \to M$ with $F(x, 0) = S_0(x)$ and $F(x, 1) = S_1(x)$. The negative curvature of (M, g_0) implies that given any $x \in M$ with $S_0(x) \neq S_1(x)$ there is a unique geodesic $t \mapsto S(x; t)$, $S(x, 0) = S_0(x)$ and $S(x, 1) = S_1(x)$ homotopic to $t \mapsto F(x, t)$ joining $S_0(x)$ and $S_1(x)$. For details see [76]. Interestingly, the proof of uniqueness involves a Morse theoretic argument on the manifold of paths joining $S_0(x)$ and $S_1(x)$.

Now let us suppose that our geodesics $S(x, t)$ have unit velocity with $S(x, 0) = S_0(x)$. Then there is a unique non-negative time $\tau(x)$ such that $S(x, \tau(x)) = S_1(x)$.
If $S_0(x) = S_1(x)$ define $\tau(x) = 0$. From the fact that for each $x \in M$ the exponential map $\exp_x : T_x M \to M$ is a local diffeomorphism it follows that τ is continuous and smooth wherever it is positive.

Let us suppose for a moment that a miracle occurs and $\tau : M \to \mathbb{R}^+$ is constant! Without loss of generality assume that $\tau \equiv 1$ and that, as before $(M, g_0) \subset \mathbb{R}^d$, isometrically. Since

$t \mapsto S(x,t)$ has unit length for all x

$$\left\|\frac{\partial S}{\partial t}\right\|^2 = 1 \tag{B.20}$$

and (B.20) holds for all $x \in M$, where $\|\cdot\|$ denotes the Euclidean $\mathbb{R}^d$ norm.

Thus the Laplace-Beltrami operator can be applied to (B.20) to yield

$$\Delta_g \left\|\frac{\partial S}{\partial t}\right\|^2 = \Delta_g \left\langle \frac{\partial S}{\partial t}, \frac{\partial S}{\partial t} \right\rangle = 0 . \tag{B.21}$$

Writing (B.21) in conformal coordinates $g_{ij} = \lambda \delta_{ij}$ we have

$$
\begin{aligned}
0 &= \tfrac{2}{\lambda} \sum_\ell \left\| \tfrac{D}{\partial x^\ell} \tfrac{\partial S}{\partial t} \right\|^2 + 2 \left\langle \tfrac{D}{\partial t} \Delta S, \tfrac{\partial S}{\partial t} \right\rangle + \tfrac{2}{\lambda} \left\langle \mathcal{R}\left(\tfrac{\partial S}{\partial x^\ell}, \tfrac{\partial S}{\partial t}\right) \tfrac{\partial S}{\partial x^\ell}, \tfrac{\partial S}{\partial t} \right\rangle \\
&= \tfrac{1}{\lambda} \sum_\ell \left\| \tfrac{D}{\partial x^\ell} \tfrac{\partial S}{\partial t} \right\|^2 + \tfrac{d}{dt} \left\langle \Delta S, \tfrac{\partial S}{\partial t} \right\rangle + \tfrac{1}{\lambda} \left\langle \mathcal{R}\left(\tfrac{\partial S}{\partial x^\ell}, \tfrac{\partial S}{\partial t}\right) \tfrac{\partial S}{\partial x^\ell}, \tfrac{\partial S}{\partial t} \right\rangle .
\end{aligned}
\tag{B.22}
$$

Hold x fixed, take the integral of (B.22) with respect to t over the interval $[0,1]$. Since

$$
\begin{aligned}
\int_0^1 \tfrac{d}{dt} \left\langle \Delta S, \tfrac{\partial S}{\partial t} \right\rangle dt &= \left\langle (\Delta S)(x,1), \tfrac{\partial S}{\partial t}(x,1) \right\rangle - \left\langle (\Delta S)(x,0), \tfrac{\partial S}{\partial t}(x,0) \right\rangle \\
&= \left\langle (\Delta S_1)(x), \tfrac{\partial S}{\partial t}(x,1) \right\rangle - \left\langle (\Delta S_0)(x), \tfrac{\partial S}{\partial t}(x,0) \right\rangle \\
&= 0
\end{aligned}
$$

we see that for each $x \in M$

$$0 = \frac{1}{\lambda} \sum_\ell \int_0^1 \left\| \frac{D}{\partial x^\ell} \frac{\partial S}{\partial t} \right\|^2 dt + \int_0^1 \frac{1}{\lambda} \left\langle \mathcal{R}\left(\frac{\partial S}{\partial x^\ell}, \frac{\partial S}{\partial t}\right) \frac{\partial S}{\partial x^\ell}, \frac{\partial S}{\partial t} \right\rangle dt .$$

But both integrals are non-negative, which implies that $\left\langle \mathcal{R}\left(\frac{\partial S}{\partial x^\ell}, \frac{\partial S}{\partial t}\right) \frac{\partial S}{\partial x^\ell}, \frac{\partial S}{\partial t} \right\rangle \equiv 0$.

Let us now use the fact that S is a diffeomorphism (to be established shortly and independently). Then this last equality implies that $\frac{\partial S}{\partial t} = 0$. This clearly contradicts the fact that $\left\|\frac{\partial S}{\partial t}\right\|^2 = 1$. Thus $S_0(x) = S_1(x)$.

If we could show now that τ is constant we would be done. This follows by applying the maximum principle and the next

Lemma B.24 $(\Delta_g \tau)(x) \geq 0$ *whenever* $\tau > 0$.

Before proving this lemma let us see how it implies that τ is constant. If $\tau(x) > 0$ for all x, then τ is a globally defined subharmonic function on M. The maximum principle states [91],[39] that τ cannot have an interior positive maximum (i.e. there cannot be a point x_0 such that $\tau(x_0) \geq \tau(x)$ for all x with strict inequality holding somewhere). Since M is compact and all points are interior points, τ must be constant and by the preceding argument $S_0(x) = S_1(x)$.

Suppose now that the set $\Gamma = \{x \mid \tau(x) = 0\} = \{x \mid S_0(x) = S_1(x)\} \neq \emptyset$. Let U be a component of $M \setminus \Gamma$. Then $\tau > 0$ on U and $\Delta_g \tau \geq 0$ on U. Again by the maximum principle for the Laplacian the maximum of τ must occur on $\partial U \subset \Gamma$. Thus $\tau \equiv 0$ on U and hence $\tau \equiv 0$ on M implying once again that $S_0(x) = S_1(x)$.

We now give the proof of lemma B.24.

PROOF OF B.24: $S(x, \tau(x)) = S_1(x)$ and $S(x, 0) = S_0(x)$. From these two equations it follows that

$$(\Delta S)(x, 0) = (\Delta S_0)(x) = 0 \tag{B.23}$$

and (in conformal coordinates)

$$0 = (\Delta S_1)(x) = \frac{1}{\lambda} \frac{D}{\partial x^\ell} \frac{\partial S_1}{\partial x^\ell} = \tag{B.24}$$

$$= \frac{1}{\lambda} \frac{D}{\partial x^\ell} \frac{\partial S}{\partial x^\ell} + \frac{2}{\lambda} \frac{D}{\partial t} \frac{\partial S}{\partial x^\ell} \cdot \frac{\partial \tau}{\partial x^\ell} + \sum_\ell \left\{ \frac{1}{\lambda} \frac{D}{\partial t} \frac{\partial S}{\partial t} \left(\frac{\partial \tau}{\partial x^\ell} \right)^2 + \frac{1}{\lambda} \frac{\partial S}{\partial t} \left(\frac{\partial^2 \tau}{\partial x^{\ell 2}} \right) \right\}$$

with the right hand side of (B.24) being evaluated at $t = \tau(x)$. However since $t \mapsto S(x, t)$ is a geodesic $\frac{D}{\partial t} \frac{\partial S}{\partial t} = 0$ and we may rewrite (B.24) as

$$(\Delta S)(x, \tau(x)) = -\frac{2}{\lambda} \frac{D}{dt} \frac{\partial S}{\partial x^\ell} \cdot \frac{\partial \tau}{\partial x^\ell} - \sum_\ell \frac{1}{\lambda} \frac{\partial S}{\partial t} \cdot \left(\frac{\partial^2 \tau}{\partial x^{\ell 2}} \right) . \tag{B.25}$$

Now equation (B.22) holds whether or not τ is constant. Let us integrate (B.22) over the interval $[0, \tau(x)]$ to obtain:

$$\sum_\ell \int_0^{\tau(x)} \frac{1}{\lambda} \left\| \frac{D}{\partial x^\ell} \frac{\partial S}{\partial t} \right\|^2 dt + \left\langle \Delta S, \frac{\partial S}{\partial t} \right\rangle \Big|_0^{\tau(x)} + \int_0^{\tau(x)} \frac{1}{\lambda} \left\langle \mathcal{R} \left(\frac{\partial S}{\partial x^\ell}, \frac{\partial S}{\partial t} \right) \frac{\partial S}{\partial x^\ell}, \frac{\partial S}{\partial t} \right\rangle dt \equiv 0 . \tag{B.26}$$

Consider the term

$$\left\langle \Delta S, \frac{\partial S}{\partial t} \right\rangle \Big|_0^{\tau(x)} = \left\langle \Delta S(x, \tau(x)), \frac{\partial S}{\partial t}(x, \tau(x)) \right\rangle - \left\langle \Delta S(x, 0), \frac{\partial S}{\partial t}(x, 0) \right\rangle . \tag{B.27}$$

But by (B.23) the second term on the right of (B.27) vanishes, and by (B.25) the first term on the right equals

$$-\frac{2}{\lambda} \left\langle \frac{D}{\partial t} \frac{\partial S}{\partial x^\ell}, \frac{\partial S}{\partial t} \right\rangle \frac{\partial \tau}{\partial x^\ell} - \sum_\ell \frac{1}{\lambda} \left\| \frac{\partial S}{\partial t} \right\|^2 \cdot \frac{\partial^2 \tau}{\partial x^{\ell 2}} .$$

But

$$\left\langle \frac{D}{\partial t}\frac{\partial S}{\partial x^\ell}, \frac{\partial S}{\partial t}\right\rangle = \left\langle \frac{D}{\partial x^\ell}\frac{\partial S}{\partial t}, \frac{\partial S}{\partial t}\right\rangle = \frac{1}{2}\frac{\partial}{\partial x^\ell}\left\|\frac{\partial S}{\partial t}\right\|^2 = 0$$

and

$$-\sum_\ell \frac{1}{\lambda}\left\|\frac{\partial S}{\partial t}\right\|^2 \cdot \frac{\partial^2 \tau}{\partial x^{\ell 2}} = -\sum_\ell \frac{1}{\lambda}\frac{\partial^2 \tau}{\partial x^{\ell 2}}\ .$$

Consequently equations (B.26) and (B.27) yield the beautifully simple equation

$$\begin{aligned}
\Delta_g \tau &= \sum_\ell \frac{1}{\lambda}\frac{\partial^2 \tau}{\partial x^{\ell 2}} \\
&= \sum_\ell \int_0^{\tau(x)}\frac{1}{\lambda}\left\|\frac{D}{\partial x^\ell}\frac{\partial S}{\partial t}\right\|^2 dt + \int_0^{\tau(x)}\frac{1}{\lambda}\left\langle \mathcal{R}\left(\frac{\partial S}{\partial x^\ell},\frac{\partial S}{\partial t}\right)\frac{\partial S}{\partial x^\ell},\frac{\partial S}{\partial t}\right\rangle dt \\
&\geq 0
\end{aligned}$$

and lemma B.24 is established.   ∎

We have therefore proved (using the existence result theorem B.23):

Theorem B.25 *For any metric g on M and any metric g_0 on M with negative scalar curvature there exists a unique C^∞ smooth harmonic map*

$$S : (M,g) \to (M,g_0)$$

homotopic to the identity.   ∎

Smooth Dependence

The vector field β on the manifold $L_3^p(M,M)$ defined by equation (B.3) actually depends smoothly on two parameters, namely the metrics g and g_0. Let us, as we did with Dirichlet's energy, consider the metric g_0 as fixed and g as a variable parameter. Then β is, in reality, a smooth function of two variables g and $u \in L_3^p(M,M)$; and we take this into account in our notation in writing $\beta(g,u)$ in place of $\beta(u)$.

If $u = S$ is a harmonic map, then the derivative of β with respect to u at S, written now as $D_u\beta(S)$ is by theorem B.2 the identity map on the tangent space $T_S L_3^p(M,M)$.

In some local coordinate system for the tangent bundle $T L_3^p(M,M)$ about the point S, $\beta(g,u) = (u, Y(g,u))$ where Y is the "principal part" of β. Then if $E = T_S L_3^p(M,M)$ we may view Y as a C^∞ map on $\mathcal{M} \times W$ into E, where $\mathcal{M}$ are all metrics g on M, W

a neighbourhood of 0 in E, with 0 corresponding to S. If S is harmonic from (M, g') to (M, g_0) then

$$Y(g', 0) = 0$$

with the derivative of Y with respect to the "second variable" u at $(g', 0)$, $D_u Y(g', 0) :$ $E \to E$ an isomorphism. In this case the implicit function theorem on Banach spaces [2] says that for some neighbourhood $W' \subset W$ and for all g sufficiently close to g', there is for each such g a unique zero $S(g)$ to $Y(g, S(g)) = 0$ and S depends smoothly on g.

Theorem B.25, already gave us global uniqueness, which is more than the local uniqueness we obtain from the implicit function theorem. However we do obtain the smooth dependence of the harmonic map S on g. Consequently we may now strengthen B.25 to

Theorem B.26 *For any metric g on M and any metric g_0 on M with negative scalar curvature there exists a unique C^∞ smooth harmonic map $S(g) : (M, g) \to (M, g_0)$ homotopic to the identity and with $g \mapsto S(g)$ being C^∞ smooth.*

The Map $S(g)$ is a Diffeomorphism

The proof that $S(g)$ is a diffeomorphism, originally due to Schoen-Yau and Sampson, is easily available in greater generality in their papers [97], [96] and in the lovely book by Jürgen Jost [57] whose presentation we follow.

We shall therefore be a bit more sketchy than in the preceding parts of this appendix, preferring to outline the main points of the proof.

Dropping the g from the notation for S we want to show that any harmonic map $S : (M, g) \to (M, g_0)$ homotopic to the identity is a diffeomorphism. In this we need only that the scalar curvature of g_0 is negative. Write the metrics g and g_0 in local conformal coordinates as

$$\lambda \, dz \, d\bar{z}$$

and

$$\rho \, dw \, d\bar{w} \ .$$

Recall that the equation for a harmonic map S (cf. equation (3.1)) can be written as

$$S_{z\bar{z}} + \frac{\rho_w}{\rho} S_z S_{\bar{z}} = 0 \ . \tag{B.28}$$

Define the functions

$$H := |\partial S|^2 = \frac{\rho}{\lambda} |S_z|^2$$

and

$$L := |\bar{\delta}S|^2 = \frac{\rho}{\lambda}|S_{\bar{z}}|^2 \ .$$

Let K_1 and K_2 be the Gauss curvatures of (M,g) and (M,g_0) respectively in conformal coordinates

$$K_1 = -\frac{2}{\lambda}\frac{\partial}{\partial z}\frac{\partial}{\partial \bar{z}}\log \lambda$$

and

$$K_2 = -\frac{2}{\lambda}\frac{\partial}{\partial z}\frac{\partial}{\partial \bar{z}}\log \rho \leq 0 \ .$$

The following lemma follows from a straightforward calculation.

Lemma B.27 *At points where H or L, respectively are non-zero we have the identities*

$$\Delta_g \log H = 2K_1 - 2K_2(H - L) \tag{B.29}$$

$$\Delta_g \log L = 2K_1 - 2K_2(H - L). \tag{B.30}$$

Therefore

$$\Delta_g \log(H/L) = -4K_2(H - L) \ . \tag{B.31}$$

PROOF: a straightforward calculation.   ∎

The quantity $H - L$ appearing in B.29 - B.31 is geometrically significant; it is the *Jacobian determinant* of the map S, which we denote by $J(S)$.

We also observe that the product

$$HL = \frac{1}{\lambda^2}\varphi\bar{\varphi}$$

where $\varphi dz^2 = \rho S_z \overline{S_{\bar{z}}}dz^2$ is a holomorphic quadratic differential on (M,g). From this it follows that if either H or L vanish on an open set they must vanish identically. We also note that in our situation H cannot vanish identically since this would imply that $J(S) \leq 0$ and therefore that $\deg S \leq 0$, contradicting the fact that $\deg S = 1$. Thus the zeros of H must be isolated.

Lemma B.28 *Suppose f is a C^1 function on (M,g) with isolated zeros such that*

$$f_{\bar{z}} = f\omega$$

where ω is C^1. Then locally $f = e^g h$ where g is C^1 and h is holomorphic.

PROOF: Find a C^1 function g with $g_z = \omega$ (locally) and set $h = e^{-g}f$. Then $h_{\bar{z}} = 0$, i.e. h is holomorphic. ∎

As a direct consequence of this and the preceding remarks we have

Theorem B.29 *Near each isolated zero z_i of H we have the expansion*

$$H = a_i|z - z_i|^{n_i} + o(|z - z_i|^{n_i}) \tag{B.32}$$

for some $a_i > 0$ and some non-negative integer n_i.

PROOF:

$$H = \frac{\rho}{\lambda}|S_z|^2 = \frac{\rho}{\lambda}S_z\overline{S_z} \ . \tag{B.33}$$

Let $f = S_z$. Then from (B.28) it follows

$$f_{\bar{z}} = f\omega$$

where $\omega = -\frac{\rho_w}{\rho}S_z$. By lemma B.28

$$S_z = e^g h \tag{B.34}$$

where h is holomorphic. Each zero of H is a zero of S_z and also of h. Since h is holomorphic it has a power series expansion about z_i. The theorem then follows from the explicit expressions (B.33) and (B.34). As a direct consequence we obtain

Theorem B.30

$$-\sum_i n_i = 2\int_M K_1 d\mu_g - 2\int_M K_2(H - L)d\mu_g \ . \tag{B.35}$$

PROOF: Integrate expression (B.29) over M with respect to the volume measure μ_g, by first deleting small discs about the zeros z_i, and integrating over M minus the union of these small discs. The right hand side of (B.29) is continuous on M and therefore the limit of the integral of the right hand side of (B.29) over $M \setminus (\bigcup \text{discs})$ as the discs shrink to 0 is clearly the right hand side of (B.35). $\log H$, however has a singularity at each z_i. Integrating $\left(\int_{M\setminus(\bigcup\text{discs})} \Delta_g \log H\right)$ by parts yields a sum of integrals about the boundaries of these small discs. Using expansion (B.32) and going to the limit we obtain the left hand side of (B.35). ∎

As a consequence of theorem B.30 we obtain

Corollary B.31 *For a harmonic map S homotopic to the identity, H > 0.*

PROOF: Consider expression (B.35). By the Gauss-Bonnet theorem

$$2 \int_M K_1 d\mu_g = 4\pi\chi(M) \,, \tag{B.36}$$

$\chi(M)$ the Euler-characteristic of M.

In the second expression we may make a change of variables $y = S(x)$. Of course, we must not and do not use here that S is a diffeomorphism. The change of variables argument needs only deg $S = 1$. Thus

$$\int_M K_2(H - L)d\mu_g = \int_M K_2 J(S)d\mu_g = \int_M K_2 d\mu_{g_0} = 2\pi\chi(M) \,, \tag{B.37}$$

the last equation following again from Gauss-Bonnet.

Inserting (B.36) and (B.37) into (B.35) we have

$$-\sum_i n_i = 0 \;.$$

Thus all $n_i = 0$ and $H > 0$.   ∎

Lemma B.32 *Let $S : (M, g) \to (M, g_0)$ be a harmonic map homotopic to the identity with $K_2 \leq 0$. Then the functional determinant*

$$J(S) = H - L \geq 0 \;.$$

PROOF: In a region where $J(S) = H - L < 0$, one would have $L > H > 0$ and therefore $\log(H/L) < 0$. In addition,

$$\Delta_g \log(H/L) = -4K_2(H - L) \leq 0$$

wherever $J(S) \leq 0$. Thus $\log(H/L)$ is superharmonic where $J(S) \leq 0$. Therefore by the maximum principle [91] for the Laplacian $\log(H/L)$ cannot have a non-positive interior minimum where $J(S) \leq 0$.

Since $\log(H/L) = 0$ on the boundary of $\{z \in M \mid J(S)(z) \leq 0\}$ and $\log(H/L) < 0$ in the interior, the set $\{z \in M \mid J(S)(z) < 0\}$ must be empty.

This allows us to prove the main result of this section.

Theorem B.33 *Let $S : (M, g) \to (M, g_0)$ be a harmonic map homotopic to the identity with $K_2 \leq 0$. Then S is a diffeomorphism.*

PROOF: We know from Lemma B.32 that $J(S) \geq 0$ on M and by corollary B.31 that $H > 0$ on M. Suppose $J(S)(z_0) = 0$. Then $H(z_0) = L(z_0) > 0$. Thus again by (B.31)

$$\Delta_g \log(H/L) = -4K_2 J(S) \tag{B.38}$$

in some open neighbourhood about z_0.

Since $J(S) \geq 0$, $K_2 \leq 0$ and $L(z_0) > 0$ there are positive constants c_1 and c_2 and a neighbourhood U of z_0 with

$$-4K_2 J(S) \leq c_1(H/L - 1) \leq c_2 \log(H/L)$$

holding in U.

Therefore applying these facts to (B.38) we have

$$\Delta_g \log(H/L) \leq c_2 \log(H/L) \tag{B.39}$$

in U.

Again applying the strong maximum principle, for elliptic second order equations [91],[39], this time to (B.39) we see that $\log(H/L)$ cannot assume a non-positive minimum in the interior of U, unless $\log(H/L)$ is constant on U. But since $H(z) \geq L(z)$ on U, $\log(H/L)(z) \geq 0$ on U and equals zero when $z = z_0$. Thus $\log(H/L) \equiv 0$ on U. This says that the set of z where $J(S)(z) = 0$ is open as well as closed. Thus if this set is non-empty it must be all of M implying that S is constant and $\deg S = 0$, which contradicts the fact that S is homotopic to the identity. ■

Putting all the results of appendix B together we arrive at the following conclusion, which we have used extensively in our development of Teichmüller theory, namely

Theorem B.34 *For any metric g on M and any metric g_0 on M with negative scalar curvature there exists a unique C^∞ smooth harmonic map $S(g) : (M, g) \to (M, g_0)$ homotopic to the identity. Moreover $S(g)$ is a diffeomorphism and $g \mapsto S(g)$ is C^∞ smooth.*

This concludes appendix B.

C   The Mumford Compactness Theorem

We prove here the compactness theorem for the moduli space (lemma 3.2.2). In its original form the theorem is due to Mumford [79]. We present another proof given by Tomi and the author [110] using only basic geometric notions instead of the uniformization theorem.

Theorem C.1 *Let M be a closed connected smooth surface, and $\{g^n\}$ be a sequence of smooth metrics of curvature -1 on M such that all their closed geodesics are bounded below in length by a fixed positive bound. Then there exist smooth diffeomorphisms f^n of M which are orientation preserving if M is oriented, such that a subsequence of $\{f^{n*} g^n\}$ converges in C^∞ towards a smooth metric.*

PROOF OF THE MUMFORD THEOREM: Since on a negatively curved surface there are no conjugate points along any geodesic, it follows that every geodesic arc is globally minimizing (with fixed end points). Therefore, any two geodesic arcs with common endpoints cannot be homotopic with fixed endpoints; otherwise, by a common Morse-theoretic argument (see Milnor [76]), there would exist a non-minimizing geodesic arc joining these endpoints.

Another way to see this is as follows: Two *homotopic* geodesic arcs on M with common endpoints would give rise to two geodesic arcs with *common* endpoints *in the universal cover* $\bar{M}$ of M, which is a negatively curved plane. If these two arcs have no interior points in common they bound a region of the type of the disc to which Gauss-Bonnet can be applied yielding a contradiction: the total curvature is negative, whereas the integral along the boundary of the region is positive. If the two arcs do have interior points in common, the same argument applies to shorter segments of these arcs.

Hence we may conclude that a lower bound ℓ on the length ℓ_n of the closed geodesics of g_n implies a bound on the injectivity radii ρ_n of $M^n = (M, g^n)$, $\rho_n \geq \rho \geq \ell/2$.

It follows on each open disc $B_R(p)$, $p \in M^n$ and $R \leq \rho$, one can introduce a geodesic polar coordinate system. By a classical result in differential geometry the metric tensor associated with g^n in these coordinates assumes the form

$$(g_{ij}^n) = \begin{bmatrix} 1 & 0 \\ 0 & f(r) \end{bmatrix}, \quad f(r) = 2\sinh^2 \frac{r}{\sqrt{2}} \tag{C.1}$$

where r denotes the polar distance.

For the area of $B_R(p)$ we obtain from (C.1) the simple estimate

$$\text{area } B_R(p) \geq \pi R^2 .$$

The genus of the manifolds M^n being fixed, the total area of M^n is determined by the Gauss-Bonnet formula if $R(g^n) = -1$. It follows that there is an upper bound, only depending on R, for the number of disjoint open discs $B_R(p)$ in M^n. Let us now take $R = \frac{1}{4}\rho$, and let $N(n)$ be the maximal number of open disjoint discs of radius $\frac{1}{4}R$ in M^n. By passing to a subsequence we can assume that $N(n) = N$ holds independently of n. It follows that, for each $n \in \mathbb{N}$, we can find points $p_j^n \in M^n, j = 1, \ldots, N$, with the property that the discs $B_{1/4R}(p_j^n)$ are disjoint while the discs $B_{1/2R}(p_j^n)$ cover M^n. Let us now denote by $\mathbb{H}$ the Poincaré upper half plane. We pick an arbitrary point $\zeta_0 \in \mathbb{H}$, e.g., $\zeta_0 = i$, the imaginary unit, and introduce geodesic polar coordinates on $B_{4R}(p_j^n) \subset M^n$ and on $B_{4R}(\zeta_0) \subset \mathbb{H}$, respectively. The corresponding metric tensors assume the same form (C.1) in each of both cases, and we may therefore conclude that there exist orientation preserving isometries

$$\varphi_j^n : B_{4R}(p_j^n) \rightarrow B_{4R}(\zeta_0), \quad \varphi_j^n(p_j^n) = \zeta_0 .$$

Let then I^n denote the set of all pairs (j, k), $1 \leq j, k \leq N$, such that

$$B_{2R}(p_j^n) \cap B_{2R}(p_k^n) \neq \emptyset .$$

By passing to a subsequence we can assume that $I^n = I$ is independent of n. For $(j, k) \in I$, the transition mappings

$$\tau_{jk}^n := \varphi_j^n \circ (\varphi_k^n)^{-1} : \varphi_k^n \left[B_{4R}(p_j^n) \cap B_{4R}(p_k^n) \right] \rightarrow \varphi_j^n \left[B_{4R}(p_j^n) \cap B_{4R}(p_k^n) \right]$$

are well-defined local isometries of $\mathbb{H}$. Before proceeding further with the proof we first want to show that any such local isometry in fact extends to a global one.

Lemma C.2 *Let $f : U \to H$ be a C^1 orientation preserving isometry on an open connected subset U of the hyperbolic plane. Then*

$$f(w) = \frac{Aw + B}{Cw + D}, \quad A, B, C, D \in \mathbb{R} \ ,$$

and $AD - BC = 1$.

PROOF: The class of maps $w \mapsto \frac{Aw+B}{Cw+D}$, $AD - BC = 1$ with real coefficients are the group of isometries of the Poincaré metric. Thus we must show that a local isometry is also a global isometry.

f is orientation preserving. Now an easy calculation shows that f must be holomorphic and has to satisfy the non-linear condition

$$|f'(w)| = \frac{\operatorname{Im} f(w)}{\operatorname{Im} w} \ . \qquad (C.2)$$

One can check that every map of the form $w \mapsto \frac{Aw+B}{Cw+D}$ as above satisfies the condition (C.2), and the set of maps from a fixed domain to itself and satisfying (C.2) form a group. Therefore, by composition with an appropriate element of the three dimensional conformal group of $\mathbb{H}$ we may assume that f satisfies the following additional conditions: f is defined in a neighbourhood of $i \in H$, and

$$f'(i) = 1, \quad \operatorname{Im} f(i) = 1 \ . \qquad (C.3)$$

Now, writing $w = u + iv$, $\partial_w = \frac{1}{2}(\partial_u - i\partial_v)$ and using (C.2), we have

$$
\begin{aligned}
(\log f')' &= 2\{\operatorname{Re}(\log f')\}_w = 2\{\log |f'|\}_w \\
&= \frac{(\operatorname{Im} f)_w}{\operatorname{Im} f} - \frac{v_w}{v} = \frac{-if'}{\operatorname{Im} f} + \frac{i}{v} \ .
\end{aligned}
$$

By (C.3), this implies $(\log f')'(i) = 0$ and hence $f''(i) = 0$. Similarly

$$(\log f')'' = -\left[\frac{if'}{\operatorname{Im} f}\right]_w + \left[\frac{i}{v}\right]_w = \frac{-if''}{\operatorname{Im} f} - \frac{if'}{(\operatorname{Im} f)^2}\left(\frac{if'}{2}\right) + \frac{i^2}{2v^2} \ . \qquad (C.4)$$

Again we see that $(\log f')'' = 0$ and hence $f'''(i) = 0$. Proceeding inductively we obtain

$$(\log f')^{(n)}(i) = 0$$

for all $n \in \mathbb{N}$. In the induction process, only the derivatives of $\operatorname{Im} f$ in the second term of the right hand side of (C.4) and those of v in the third term contribute, but these contributions cancel. All other contributions vanish by the inductive hypothesis

$f''(i) = \ldots = f^{(n)}(i) = 0$. Thus, since f is holomorphic in a neighbourhood of i, it follows that f' is constant; therefore $f(w) = w$. Since we normalized f by the isometry group of $\mathbb{H}$, this proves that our initial map f must be in this isometry group, and the proof of the lemma is complete. ∎

For $(j, k) \in I$ we have $p_k^n \in B_{4R}(p_j^n)$, and hence $q_{jk}^n := \varphi_j^n(p_k^n) \in B_{4R}(\zeta_0)$, since φ_j^n is an isometry. It is obvious from the definition that

$$q_{jk}^n = \tau_{jk}^n(q_{kj}^n) \ . \tag{C.5}$$

We are now going to construct a limit manifold of the sequence $M^n = (M, g^n)$. For this purpose we prove

Lemma C.3 *The family of transition mappings* $(\tau_{jk}^n)_{n \in \mathbb{N}}$ *is compact for each* $(j, k) \in I$.

PROOF: By Lemma C.2, each τ_{jk}^n is a global isometry of $\mathbb{H}$ and there is a fixed compact subset K of $\mathbb{H}$ and points $q_{jk}^n \in K$ such that (C.5) holds. By composition with a conformal map of $\mathbb{H}$ onto the unit disc $\mathcal{B} \subset \mathbb{R}^2$ we may assume that each τ_{jk}^n is a conformal map of $\mathcal{B}$ onto itself and (suppressing the indices j, k) that there are points p^n strictly staying away from $\partial \mathcal{B}$ such that also $\tau^n(p^n)$ stays uniformly away from $\partial \mathcal{B}$. Each τ^n is of the form

$$\tau^n(w) = d_n \frac{w - a_n}{1 - \bar{a}_n w} \ ,$$

where $|a_n| < 1$, $|d_n| = 1$. It suffices to show that $|a_n|$ stays strictly below 1. If not, we can assume $a_n \to a$, $|a| = 1$, and $d_n \to d$, $|d| = 1$. The limit map then

$$\tau(w) = d \frac{w - a}{1 - \bar{a} w} = a d \left\{ \frac{\bar{a} w - 1}{1 - \bar{a} w} \right\} = -ad$$

collapses the disc onto a point on $\partial \mathcal{B}$ which is a contradiction. ∎

We can now continue with the proof of Mumford's theorem. Passing to a subsequence we can by Lemma C.3 assume that

$$\tau_{jk}^n \to \tau_{jk} \quad \text{as } n \to \infty \ . \tag{C.6}$$

We now define a limiting manifold $\hat{M}$ as the disjoint union of N discs $B_R(\zeta_0) \subset H$, labelled as $B_1, \ldots, B_N$ with the identifications

$$p \in B_j \text{ equals } q \in B_k \quad \Leftrightarrow \quad (j, k) \in I \text{ and } p = \tau_{jk}(q) \ .$$

It is clear that $\hat{M}$ is a differentiable manifold carrying a natural Riemannian metric which on each B_j coincides with the Poincaré metric. We claim that $\hat{M}$ is compact. Assume to the contrary that there were a point $q \in \partial B_R(\zeta_0)$ such that $q \notin \tau_{jk}(B_R(\zeta_0))$ for some j and all k with $(j,k) \in I$. Then it would follow that, for sufficiently large n, we have $q \notin \tau_{jk}^n\left[B_{(3/4)R}(\zeta_0)\right]$ which means that $(\varphi_j^n)^{-1}(q) \notin B_{(3/4)R}(p_k^n)$. This, however, would imply that

$$B_{(1/4)R}\left[(\varphi_j^n)^{-1}(q)\right] \cap B_{(1/4)R}(p_k^n) = \emptyset \quad \text{for} \quad k = 1, \ldots, N \ ,$$

contradicting the choice of N as the maximal number of disjoint discs in M^n of radius $(1/4)R$. The remainder of the proof rests upon the following

Lemma C.4 *There are diffeomorphisms $f^n : \hat{M} \to M^n$, $f^n(B_j) \subset B_{2R}(p_j^n)$ such that*

$$\varphi_j^n \circ f^n \to id \quad \text{in } C^\infty \quad \text{on each } B_j, \quad \text{as } n \to \infty \ . \tag{C.7}$$

The proof of this lemma is somewhat technical and presented below.

Let us first quickly finish the proof of Mumford's theorem assuming the lemma. Denoting by g the Poincaré metric, we have from (C.7) that

$$f^{n*}\varphi_j^{n*}g \to g \quad \text{as } n \to \infty$$

on each B_j. Since, however, φ_j^n was an isometry between g and g^n on M^n, this means that

$$f^{n*}g^n \to g \quad \text{as } n \to \infty$$

on $\hat{M}$. Choosing now any (symmetric) diffeomorphism $f : M \to \hat{M}$, we obtain

$$(f^n \circ f)^*g^n \to f^*g \quad \text{as } (n \to \infty) \ ,$$

which proves Mumford's theorem.   ■

We now come to the proof of Lemma C.4.

For the proof let us consider the manifold M^n as the disjoint union of N-balls $B_{2R}(\zeta_0) \subset H$ labelled as $B_1', \ldots, B_N'$ with the identifications

$$x \in B_i' \quad \text{equals} \quad y \in B_j' \quad \text{iff} \quad (i,j) \in I \quad \text{and} \quad x = \tau_{ij}^n(y) \ .$$

We denote this model of M^n by $\hat{M}^n$. It then suffices to show that there are diffeomorphisms $f^n : \hat{M} \to \hat{M}^n$, $f^n(B_i) \subset B_i'$, such that $f_n \to id$ (as $n \to \infty$) on each B_i. We shall do this by a Morse theoretic argument.

Since $\hat{M}$ is an oriented closed surface there is a C^∞ Morse function $\psi : \hat{M} \to \mathbb{R}$ with distinct critical values $c_0 > c_1 > \ldots > c_m$ and such that the level sets $\psi^{-1}(c_j)$ contain only one non-degenerate critical point w_j.

We may use a partition of unity to construct a sequence of functions $\psi^n : M^n \to \mathbb{R}$ such that on each B_i, $\psi^n \to \psi$ in C^∞. To see this let φ_i be the natural coordinate charts on $\hat{M}$ induced by the inclusion of B_i into $\mathbb{H}$, so that $\varphi_i \circ \varphi_j^{-1} = \tau_{ij}$. Furthermore, let $\{\eta_i\}$ be a partition of unity on $\hat{M}$ with respect to the coordinate cover $\{B_i\}$. Define $\psi^n : M^n \to \mathbb{R}$ by

$$\psi^n(p) = \sum_j \eta_j \left(\varphi_j^{-1} \varphi_j^n(p) \right) \psi \left(\varphi_j^{-1} \varphi_j^n(p) \right) \ .$$

If $p = (\varphi_k^n)^{-1}(u)$ then

$$\psi^n(p) = \sum_j \eta_j \left(\varphi_j^{-1} \tau_{ij}^n(u) \right) \psi \left(\varphi_j^{-1} \tau_{jk}^n(u) \right) \ .$$

As $n \to \infty$ this converges C^∞ to

$$\sum_j \eta_j \left(\varphi_j^{-1} \tau_{jk}(u) \right) \psi \left(\varphi_j^{-1} \tau_{jk}(u) \right) = \sum_j \eta_j \left(\varphi_k^{-1}(u) \right) \psi \left(\varphi_k^{-1}(u) \right) = \psi \left(\varphi_k^{-1}(u) \right) \ ,$$

which (after viewing ψ^n as defined on $\hat{M}^n$) proves the result.

Consequently, for large n, ψ^n has non-degenerate critical points $\{w_j^n\}$ "near" the $\{w_j\}$ on $\overset{N}{\underset{\ell=1}{\bigcup}} B_\ell$. By trivial modifications of ψ^n we may further assume that ψ^n has the same critical values $c_0 > \ldots > c_m$ as does ψ and that $w_j^n = w_j$ for all j.

Furthermore we can assume that in a small disc about each w_j (in the B_i's) ψ_n and ψ actually agree.

Let $(\hat{M}^n)^a = \{x | \psi^n(x) \le a\}$ and $(\hat{M}^n)_a = \{x | \psi^n(x) \ge a)\}$ with $(\hat{M})^a$, $(\hat{M})_a$ defined similarly in terms of ψ. Let $\varepsilon > 0$ be small enough so that $(\hat{M}^n)_{c_0 - 2\epsilon}$ and $(\hat{M})_{c_0 - 2\epsilon}$ contain only w_0 as its only critical point and $(\hat{M}^n)_{c_1 - 2\epsilon}$ and $(\hat{M})_{c_1 - 2\epsilon}$ contain only w_0 and w_1.

Let G be a fixed metric on $\hat{M}$ which agrees with the Euclidean metric on a neighbourhood of the $\{w_j\}$. As in constructing the ψ^n we can easily find a sequence of metrics G_n on $\hat{M}^n$ so that G_n agrees with G on a neighbourhood of the (w_j) (in $\bigcup B_i$) and $G_n \to G$ as $n \to \infty$. Let $\nabla \psi^n$ and $\nabla \psi$ denote the gradients of ψ^n and ψ with respect to these metrics, and X^n and X the normalized fields $\frac{\nabla \psi^n}{\|\nabla \psi^n\|}$ and $\frac{\nabla \psi}{\|\nabla \psi\|}$, $\|\cdot\|$ denoting the norms with respect to G_n and G. Of course X^n and X are defined only on $\hat{M}^n - \bigcup w_j$ and $M - \bigcup w_j$

respectively. We shall define a mapping $f_n : (\hat{M}_n)_{c_1-2\epsilon} \to \hat{M}$ which is a diffeomorphism of a neighbourhood of $(\hat{M}_n)_{c_1-\epsilon}$ to a neighbourhood of $(\hat{M})_{c_1-\epsilon}$.

Let D_0 be a "small disc" about w_0 for which the Morse lemma holds for ψ^n and ψ about w_0. Thus there exists a map Q from a neighbourhood of 0 in $\mathbb{R}^2$ to a neighborhood of w_0 so that $\psi^n \circ Q(z) = c_0 - z_1^2 - z_2^2 = \psi \circ Q(z)$. Thus we may take $D_0 = \{x | \psi_n(x) \geq c_0 - \epsilon\}$. By the Morse lemma it follows that both $(\hat{M})_{c_0-\epsilon}$ and $(\hat{M}^n)_{c_0-\epsilon}$ are diffeomorphic to a ball and hence to each other. The idea of our proof is to now proceed down the critical points to show that $(\hat{M}^n)_{c_j-\epsilon}$ is diffeomorphic to $(\hat{M})_{c_j-\epsilon}$ for $j = 1, \cdots, m$. It is enough to indicate how this is done for $j = 1$. Let $p \in \partial D_0$, and let $\sigma_p^n(t)$ and $\sigma_p(t)$ be the flows of the vector fields X^n and X respectively with $\sigma_p^n(0) = p = \sigma_p(0)$. It follows immediately that $\psi^n\left(\sigma_p^n(t)\right) = c_0 - \epsilon + t = \psi\left(\sigma_p(t)\right)$ and that $\sigma_p^n(t)$ (resp. $\sigma_p(t)$) as t decreases either converges to w_1 or drops into $(\hat{M}^n)^{c_1-2\epsilon}$ (resp. $(\hat{M})^{c_1-2\epsilon}$). Let U be the unstable manifold of w_1 for the flow of X^n. Then it follows that every $q \in (\hat{M}_n)_{c_1-2\epsilon} \backslash (U \cup w_0)$ can be written as $\sigma_p(t)$ for some $t \in \mathbb{R}$ and $p \in \partial D_0$. Define the map

$$\tilde{f}^n : (\hat{M}^n)_{c_1-2\epsilon} \backslash U \to \hat{M}$$

by $\tilde{f}^n\left(\sigma_p^n(t)\right) = \sigma_p(t)$, $p \in \partial D_0$ and $\tilde{f}^n(w_0) = w_0$. Since X^n and X agree on D_0 (in some coordinate system) it follows that f_n is the identity in a neighbourhood of w_0, with respect to the above coordinate system, and is thus smooth everywhere it is defined. It also follows from our construction that

$$\psi^n(w) = \psi\left(\tilde{f}^n(w)\right)$$

and so $\tilde{f}^n$ takes level sets to level sets and also that $\tilde{f}^n \to id$ as $n \to \infty$ (on $\bigcup B_j$).

Now let us assume that we are in a coordinate neighbourhood W_1 of w_1 where Morse's lemma holds for ψ^n and ψ and where $\psi^n \equiv \psi$. The situation is as depicted in the figure below.

Let D_1' and $D_1, \bar{D}_1' \subset D_1$ be two strips as in figure C.1. Let η be a C^∞ function which is 1 on $W_1 \backslash D_1$ and 0 on D_1'. Define a new map $f^n : (\hat{M}^n)_{c_1-2\epsilon} \to \hat{M}$ by

$$f^n = \eta \tilde{f}_n + (1-\eta)id \ .$$

It is clear that for sufficiently large n, f_n is a diffeomorphism. Taking now the initial values of our trajectories to lie on $(\psi^n)^{-1}(c_1 - 2\epsilon)$ and $\psi^{-1}(c_1 - 2\epsilon)$, we can proceed inductively to extend f_n to a diffeomorphism of $\hat{M}^n$ onto $\hat{M}$. This completes the proof of Theorem C.1.

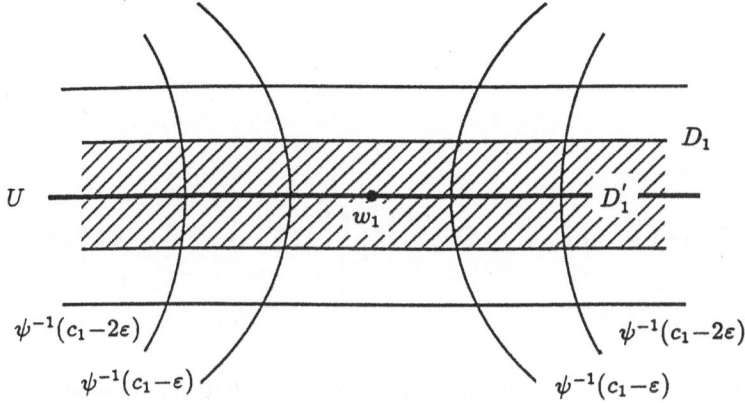

Figure C.1:

D Proof of the Collar Lemma

A stronger version of the collar lemma can be found in [92]. The original references are [48] and [59]. The weaker result given here does not give a lower bound for the area of the collar as $\ell \to 0$ in contrast to [92]; its advantage is that it avoids some topology.

Let us repeat here the lemma 3.2.1 to be proved, except for a slight change of notation and a rescaling which changes $R \equiv -1$ $\left(K \equiv -\frac{1}{2}\right)$ to $K \equiv -1$. This corresponds to changing all lengths ℓ into $\ell \cdot \sqrt{2}$:

Lemma D.1 *Let α be a (non-trivial) closed geodesic of length ℓ on a surface (M, g) where $g \in \mathcal{M}_{-1}$. Then there exists a neighbourhood U of α in M which is isometric to the following set $T/\sim$ in the hyperbolic plane: $T = \left\{(r, \theta) \mid 1 \le r \le e^\ell,\ \theta_0 \le \theta \le \pi - \theta_0\right\}$ and $\sim$ identifying $(1, \theta)$ with (e^ℓ, θ). Here, ℓ and θ_0 satisfy the estimate*

$$\cot^4 \frac{\theta_0}{2} \ge \frac{1 + \cosh \ell}{\sinh \ell} \ .$$

PROOF: It is obvious that *some* collar can be put around the closed geodesic. What is not trivial is the bound for θ_0. Therefore let $\tilde{\theta}_0$ be the infimum of all θ_0 such that the collar given in the hyperbolic plane projects injectively onto the manifold. We are going to work with this $\tilde{\theta}_0$ exclusively and drop the tilde again. This gives us the following picture: The segment from i to ie^ℓ in $\mathbb{H}$ has length ℓ and projects to the closed geodesic α on M. The geodesics in $\mathbb{H}$ orthogonal to the imaginary axis do not intersect at all in $\mathbb{H}$, and the projections on M of their parts inside T do not intersect either due to the choice of θ_0. But there are two points $\tilde{Q}_1, \tilde{Q}_2$ on the boundary of the collar (i.e. $\theta = \theta_0$ or $\theta = \pi - \theta_0$, and $1 \le r \le e^\ell$) which project to the same point Q on M. Let $\tilde{P}_1, \tilde{P}_2$ be their foot points of their respective (hyperbolic) perpendiculars onto the imaginary axis. The two perpendiculars together project to a geodesic segment $[P_1 Q P_2]$ on M. There is no "angle" at Q in this segment, for else a shorter segment $[P_{1'} P_{2'}]$ from α to α in the same homotopy class on M could be found (*some* segment by cutting short at Q, and an even shorter *geodesic* segment by minimizing the length in this homotopy class). Such a shortest geodesic segment would also have to be perpendicular on α by the same short-cut argument and would hence arise from the same construction as did $[P_{1'}P_{2'}]$. This would contradict the choice of θ_0. We are looking for a lower bound for the length of the segment $[P_{1'}P_{2'}]$.

We do not know whether $\tilde{Q}_1$ and $\tilde{Q}_2$ are on the same component of the boundary of the collar, $\theta(\tilde{Q}_1) = \theta(\tilde{Q}_2)$, or on different components, $\theta(\tilde{Q}_1) = \pi - \theta(\tilde{Q}_2)$. Moreover we cannot guarantee that $P_1 = P_2$. This latter property can however be achieved by doubling the surface M: cut M along α and glue two copies together thus obtaining a new manifold, $2M$.

We have now two closed geodesics α and $\beta = P_1 Q P_2 Q' P_1$ intersecting orthogonally at P_1 on $2M$. The length of α is ℓ. By construction, for any point on β, the shortest (on M) perpendicular to α is a segment of β. Let us consider this situation in the universal cover: α can be lifted to a geodesic segment $\tilde{\alpha}$ from $\tilde{A}$, i.e. $ie^{-\ell}$ to $\tilde{B} = i$. Then β can be lifted to a geodesic segment from $\tilde{B}$ to $\tilde{C}$. Continuing this way, the path $\alpha\beta\alpha^{-1}\beta^{-1}$ on $2M$ can be lifted to a path of segments with corners at $\tilde{A}, \tilde{B}, \tilde{C}, \tilde{D}, \tilde{E}$ all of which project to P_1. We claim that the hyperbolic lines $\tilde{D}\tilde{E}$ and $\tilde{A}\tilde{B}$ do not intersect. As we shall prove quantitatively later, this is the property which bounds the length of β. But the reader can immediately glance at the figure to see that for *short* β the lines *would* intersect.

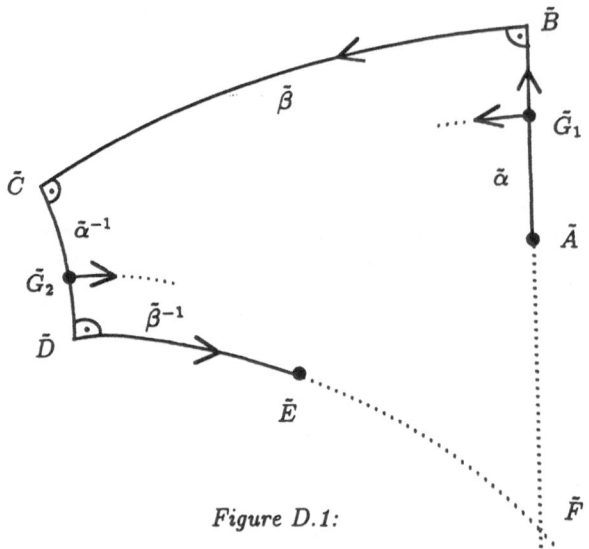

Figure D.1:

So suppose now that $\tilde{D}\tilde{E}$ and $\tilde{A}\tilde{B}$ intersect in $\tilde{F}$ (between $\tilde{A}$ and $\tilde{B}$ or not, and between $\tilde{D}$ and $\tilde{E}$ or not, as the case may be). The divergence of geodesics due to negative curvature actually rules out the possibility that $\tilde{F}$ is between $\tilde{A}$ and $\tilde{B}$. For constant curvature this calculation needs only elementary hyperbolic geometry and will be given below. We intend to show under this assumption that the projection of the quadrangle $\tilde{F}\tilde{B}\tilde{C}\tilde{D}$ covers all of $2M$ and derive a contradiction from this. To this end let $P \in 2M$ be any point not on β. Drop a shortest perpendicular γ from P to α (its footpoint being G) and construct a lift $\tilde{\gamma}$ of γ as follows:

Choose $\tilde{G}$ over G on the segment $[\tilde{A}\tilde{B}]$ if this makes $\angle(\tilde{G}\tilde{B}, \tilde{G}\tilde{P})$ a positively oriented angle. Else choose $\tilde{G}$ over G on the segment $[\tilde{C}\tilde{D}]$.

We have one of the two situations sketched in figure D.1. γ being a shortest perpendicular to α, it cannot intersect β (unless it is contained in β, which we excluded). Therefore the line $\tilde{G}\tilde{P}$ cannot intersect either of the lines $\tilde{B}\tilde{C}$ or $\tilde{F}\tilde{D}$. Nor can γ intersect α except at its endpoint G. Therefore the segment $[\tilde{G}\tilde{P}]$ cannot intersect either of the lines $\tilde{C}\tilde{D}$ or $\tilde{A}\tilde{B}$. Thus $\tilde{P}$ must be inside the quadrangle $\tilde{F}\tilde{B}\tilde{C}\tilde{D}$ whose projection therefore covers all of $2M$. Its area is $2\pi - 3 \cdot \frac{\pi}{2} - (\text{angle at } \tilde{F}) < \frac{\pi}{2}$. But the area of $2M$ is $2 \cdot 2\pi(2g-2) \geq 8\pi$, so we get a contradiction that leaves us with the conclusion that $\tilde{D}\tilde{E}$ and $\tilde{A}\tilde{B}$ do not intersect.

Now look at formula (D.1) and figure D.2 below. It not only proves $\ell' > \ell$, hence $\tilde{F}$ is not between $\tilde{A}$ and $\tilde{B}$, but it also gives the condition for $\tilde{F}$ not to exist which is our actual situation. It is $\sinh\ell\sinh d \geq 1$. Now d is twice the length of β, i.e. $4\ln\cot\frac{\theta_0}{2}$. This immediately gives the bound and leaves us only with the task of providing the details of hyperbolic trigonometry used for deriving formula (D.1). One makes use of the law of sines and the two laws of cosines in hyperbolic trigonometry.

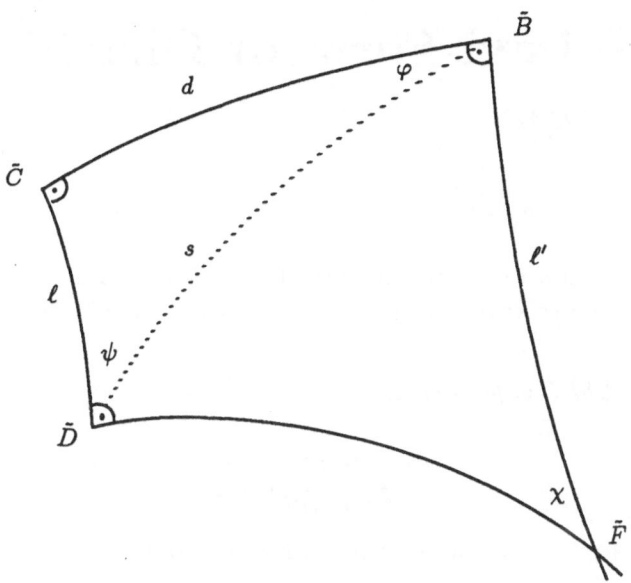

Figure D.2:

law of cosines, $\triangle\tilde{B}\tilde{C}\tilde{D}$	$\cosh s = \cosh d \cosh \ell$
law of sines, $\triangle\tilde{B}\tilde{C}\tilde{D}$	$\dfrac{\sin\varphi}{\sinh\ell} = \dfrac{\sin\psi}{\sinh d} = \dfrac{1}{\sinh s}$
law of cosines, $\triangle\tilde{D}\tilde{F}\tilde{B}$	$\cos\chi = -\sin\varphi\sin\psi + \cos\varphi\cos\psi\cosh s$
law of sines, $\triangle\tilde{D}\tilde{F}\tilde{B}$	$\dfrac{\sinh\ell'}{\cos\psi} = \dfrac{\sinh s}{\sin\chi}$

Calculate $\sinh^2\ell'$ from the last of these equations in terms of d and ℓ using the first three equations to eliminate the other variables: the horrible formula simplifies considerably by noting that $\cosh^2 d\cosh^2\ell - 1 - \sinh^2 d = \cosh^2 d\sinh^2\ell$.

The result is

$$\frac{\sinh\ell'}{\sinh\ell} = \frac{\cosh d}{\sqrt{1 - \sinh^2\ell\sinh^2 d}} > \cosh d > 1 \ . \tag{D.1}$$

This ends the proof of our version of the lemma.

E The Levi-Form of Dirichlet's Energy

We present here a direct computation of the Levi-form of Dirichlet's energy $\tilde{E}$ on $T(M)$ using the Abresch-Fischer coordinates introduced in section 4.3. The result is:

Theorem E.1 *The Levi-form of $\tilde{E}$,*

$$\sum_{\alpha\beta} \frac{\partial^2 \tilde{E}}{\partial z^\alpha \partial \bar{z}^\beta}[g]\xi^\alpha \bar{\xi}^\beta \tag{E.1}$$

where $\xi^\alpha = \gamma^\alpha + i\rho^\alpha$ is given in the Abresch-Fischer coordinates by

$$
\begin{aligned}
\frac{\partial^2 \tilde{E}}{\partial z^\alpha \partial \bar{z}^\beta}[g]\xi^\alpha \bar{\xi}^\beta \; = \; & \frac{1}{2}\sum_\ell \int_M \{h \cdot h\} g(x)(\nabla_g S^\ell, \nabla_g S^\ell) d\mu_g \tag{E.2} \\
& - \sum_r \int_M \left\{ \left\| \nabla_{\frac{\partial}{\partial x^r}} w^h \right\|^2 + \left\| \nabla_{\frac{\partial}{\partial x^r}} w^{ih} \right\|^2 \right\} dx \, dy \\
& + \sum_r \int_M g_0(x) \left(\mathcal{R}\left(\frac{\partial S}{\partial x^r}, w^h \right) w^h, \frac{\partial S}{\partial x^r} \right) dx \, dy \\
& + \sum_r \int_M g_0(x) \left(\mathcal{R}\left(\frac{\partial S}{\partial x^r}, w^{ih} \right) w^{ih}, \frac{\partial S}{\partial x^r} \right) dx \, dy
\end{aligned}
$$

where $(x^1, x^2) = (x, y)$ and h is the horizontal lift of $h^* = \sum_\alpha \xi^\alpha h_\alpha^*$, h_α^* a basis for $T_{[g]}T(M)$ over $\mathbb{C}$ defining the local coordinates z^α, z^β. Moreover $w^h = DS(g)h$, $DS(g)$ the derivative of S with respect to g in the direction h, $\mathcal{R}$ the curvature tensor of (M, g_0) and where i denotes the multiplication by i for the complex structure on $T(M)$ induced by the complex structure on $\mathcal{A}$. Finally $\nabla_{\frac{\partial}{\partial x^r}}$ denotes covariant differentiation with respect to g_0 "along S".

PROOF: We must compute

$$D^2 \tilde{E}[g](h^*, h^*) + D^2 \tilde{E}[g](ih^*, ih^*) \tag{E.3}$$

for any $h^* \in T_{[g]}(\mathcal{M}_{-1}/D_0), h^* = \sum \xi^\alpha h_\alpha^*$.

For g a Riemannian metric on M, let $\hat{E}(g) = E_g(S(g))$. Let $\tilde{\Psi} : \mathcal{A} \to \mathcal{M}_{-1}$ be the map which assigns the Poincaré metric g to $\xi \in \mathcal{A}$ and ψ a complex coordinate system for $\mathcal{A}$ about $J_0 = \tilde{\Psi}^{-1}(g)$. Therefore (E.3) is equivalent to computing

$$D^2(\hat{E} \circ \varphi)(0)(H, H) + D^2(\hat{E} \circ \varphi)(0)(J_0 H, J_0 H) \tag{E.4}$$

for all $H \in \mathcal{H}^{TT}(J_0)$ where $\varphi = \tilde{\Psi} \circ \psi$ and the derivatives are computed at $0 \in T_{J_0}\mathcal{A}$.

$$D^2(\hat{E} \circ \varphi) = D^2\hat{E}(D\varphi, D\varphi) + D\hat{E} \circ D^2\varphi \ . \tag{E.5}$$

We would like to compute $D\varphi(0)H$ and $D^2\varphi(0)(H, H)$, the first and second derivatives of φ at the origin and evaluated at H and (H, H) respectively. Let S_2 be the space of symmetric C^∞ 0-2 tensors and $S_2^{TT}(g)$ denote the trace for divergence free symmetric two tensors with respect to g. Then from (2.6) we know that $D\tilde{\Psi}(J) : T_J\mathcal{A} \to T_g\mathcal{M}_{-1} \subset S_2$ is given by

$$D\tilde{\Psi}(J)\dot{J} = \rho g + h \tag{E.6}$$

where $g = \tilde{\Psi}(J), h = -(J\dot{J})_\flat$ and

$$\Delta\rho - \rho = \delta_g\delta_g h \ ,$$

Δ the Laplace-Beltrami operator on functions.

Let $L_g = \Delta - I, I$ the identity. Then

$$\rho = L_g^{-1}(\delta_g\delta_g h) \ .$$

If h is divergence free then $\rho = 0$. From equation (4.9) it follows that

(i) $D\psi(H)(\dot{J}_1) = \dot{J}_1 J_0(I + H)^{-1} - (I + H)J_0(I + H)^{-1}\dot{J}_1(I + H)^{-1}$

(ii) $D\psi(0)(\dot{J}_1) = -2J_0\dot{J}_1$

(iii) $D^2\psi(0)(\dot{J}_1, \dot{J}_2) = 2J_0(\dot{J}_1\dot{J}_2 + \dot{J}_2\dot{J}_1)$.

Therefore

$$\begin{aligned} D\varphi(H)\dot{J}_1 &= D\tilde{\Psi}(J) \circ D\psi(H)\dot{J}_1 \\ &= \left(-J\, D\psi(H)\dot{J}_1\right)_\flat + \rho g \\ &= \left(-J\, D\psi(H)\dot{J}_1\right)_\flat + \rho(J) \cdot \tilde{\Psi}(J) \ , \end{aligned} \tag{E.7}$$

$g = \tilde{\Psi}(J), J = (I+H)J_0(I+H)^{-1}$ and $\rho(J) = L_g^{-1}\left(\delta_g \delta_g\left(-J\, D\psi(H)\dot{J}_1\right)\right)$ where, as usual $\flat$ denotes lowering an index via the metric g.

Now $\psi(0) = J_0$ and $D\psi_0(\dot{J}_1)$ is a trace free divergence free tensor, whence it follows that $\rho(J_0) = 0$.

Let us first consider the term

$$H \mapsto \left(-J\, D\psi(H)\dot{J}_1\right)_\flat$$

in expression (E.7) for which we would like to compute the derivative in the direction J_2. But

$$
\begin{aligned}
\left(-J\, D\psi(H)\dot{J}_1\right)_\flat &= -\left((I+H)J_0(I+H)^{-1}D\psi(H)\dot{J}_1\right)_\flat \\
&= -\left((I+H)J_0(I+H)^{-1}\dot{J}_1 J_0(I+H)^{-1} + \dot{J}_1(I+H)^{-1}\right)_\flat.
\end{aligned}
$$

For $H = 0$ we obtain $-2\dot{J}_1 = D\psi(0)\dot{J}_1$. The derivative of

$$H \mapsto -\left((I+H)J_0(I+H)^{-1}\dot{J}_1 J_0(I+H)^{-1} + \dot{J}_1(I+H)^{-1}\right)$$

at 0 in the direction of $\dot{J}_2$ is easily computed to be

$$2(\dot{J}_1\dot{J}_2 - \dot{J}_2\dot{J}_1)\ . \tag{E.8}$$

Consider now the map

$$H \mapsto \tilde{\Psi}(J)_{i\ell}A_j^\ell = (A_j^\ell)_\flat \tag{E.9}$$

where A is a fixed 1-1 tensor. The derivative of this at 0 in the direction $\dot{J}_2$ is

$$\left(D\tilde{\Psi}(J_0)D\psi(0)\dot{J}_2\right)_{i\ell}A_j^\ell = \left((-2\dot{J}_2)_\flat\right)_{i\ell}A_j^\ell\ . \tag{E.10}$$

In the case $A_j^\ell = -2\dot{J}_1$ we see that this is equal to

$$= 4(\dot{J}_2\dot{J}_1)_\flat\ .$$

Adding this and (E.8) together we find that the derivative of

$$H \mapsto \left(-J\, D\psi(H)\dot{J}_1\right)_\flat$$

at 0 is the bilinear map

$$(\dot{J}_1, \dot{J}_2) \mapsto 2(\dot{J}_1\dot{J}_2 + \dot{J}_2\dot{J}_1)_\flat\ .$$

Thus in order to complete our computation of the derivative of

$$H \mapsto D\varphi(H)\dot{J}_1$$

we must consider the second term in the final expression (E.7) on the derivative of the map

$$J \mapsto \rho(J)\tilde{\Psi}(J)$$

at the point J_0. Since $\rho(J_0) = 0$ we need only calculate $D\rho(J_0)\dot{J}_2$. Let $X = J\,D\psi(0)\dot{J}_1$, $Y = D\psi(0)\dot{J}_2$ and $g = \tilde{\Psi}(J_0)$. Then since X and Y are trace free divergence free it follows that

$$D\rho(J_0)Y = L_g^{-1}\left(\delta_g D_g \delta_g(Y)X\right) \tag{E.11}$$

where $D_g\delta_g(Y)$ is the derivative of the divergence operator δ_g with respect to g in the direction Y. Thus we have our formula for $D^2\varphi(0)$, namely

$$D^2\varphi(0)(\dot{J}_1, \dot{J}_2) = 2(\dot{J}_1\dot{J}_2 + \dot{J}_2\dot{J}_1)_\flat + L_g^{-1}\left(\delta_g(D_g\delta_g)(Y)X\right) \ ,$$

where $X = J\,D\psi(0)\dot{J}_1, Y = J\,D\psi(0)\dot{J}_2$.

Lemma E.2

$$\delta_g(D_g\delta_g)(X)X = 0 \ .$$

PROOF: Since $(D_g\delta_g)(Y)X = -\delta_g D_X Y$, $D_X Y$ the derivative of Y with respect to X, then by (5.15) it follows that

$$(D_g\delta_g)(X)X = -\frac{1}{2}*d\mu \ ,$$

μ a real valued function on M, and $*d\mu$ the Hodge dual. Thus

$$\delta_g\left(D_g\delta_g(X)X\right) = -\frac{1}{2}\delta_g(*d\mu) = 0 \ .$$

This gives us

Theorem E.3

$$D^2\varphi(0)(H, H) = 4(H^2)_\flat \ .$$

We are now ready to complete the proof of Theorem E.1. By formula (E.5) we must compute the sum of $D^2\hat{E}(D\varphi(0)H, D\varphi_0 H)$, $D\hat{E}\circ D^2\varphi(0)(H, H)$, $D^2\hat{E}(D\varphi(0)J_0 H, D\varphi(0)J_0 H)$, and $D\hat{E}\cdot D^2\varphi(0)(J_0 H, J_0 H)$.

Now for $h \in S_2^{TT}(g), h^* \in T_{[g]}\mathcal{T}(M), D\hat{E}(g)(h) = D\bar{E}[g]h$, by lemma 3.1.4 we see that for k arbitrary

$$D\hat{E}(g)k = -\frac{1}{2}\sum_\ell \int_M g(x)(K^T\nabla_g s^\ell, \nabla_g s^\ell)d\mu_g$$

where $K = (k)^{\natural}$ and K^{τ} is the trace free part of K. Therefore

$$D\hat{E}(g)D^2\varphi(0)(H, H) = -2\sum_{\ell}\int_M g(x)\left((H^2)^{\tau}\nabla S^{\ell}, \nabla S^{\ell}\right) d\mu_g .$$

Lemma E.4 *If* $H \in T_J\mathcal{A}$ *is divergence free then* $H^2 = \mu I$ *where* μ *is a non-negative function which vanishes at most at finitely many points of* M.

PROOF: Write H in conformal coordinates $g_{ij} = \lambda\delta_{ij}$ as $H = \begin{pmatrix} a & b \\ b & -a \end{pmatrix}$. Then $\lambda a - i\lambda h$ is a holomorphic quadratic differential on M and thus as 4(genus M)-4 zeros (genus $M >$ 1). $H^2 = (a^2 + b^2)I = \mu I, \mu = \frac{1}{2}$ trace H^2, which concludes the proof of the lemma. Consequently we see that

$$D\hat{E}(g)\left(D\varphi(0)(H), D\varphi(0)(H)\right) = D^2(\hat{E}\circ\varphi)(0)(H, H) .$$

If $h = D\varphi(0)(H) = (-2H)_{\natural}$ then by formula 3.5

$$D^2\hat{E}(g)(h, h) = \frac{1}{4}\sum_{\ell}\int_M (h\cdot h)g(x)(\nabla_g S^{\ell}, \nabla_g S^{\ell})d\mu_g - D^2 E_g(S)(w^h, w^h) . \qquad (E.12)$$

Therefore

$$D^2(\hat{E}\circ\varphi)(H, H) + D^2(\hat{E}\circ\varphi)(J_0 H, J_0 H) =$$
$$\frac{1}{2}\sum_{\ell}\int_M \{h\cdot h\}g(x)(\nabla_g S^{\ell}, \nabla_g S^{\ell})d\mu_g - D^2 E_g(S)(w^h, w^h) - D^2 E_g(S)(w^{ih}, w^i h)$$

since if $ih = (-2J_0 H)_{\natural}$, then $\{ih\cdot ih\} = \{h\cdot h\}$.

As we know the second variation of E_g, namely

$$D^2 E_g(S)(w, w) = \int_M \left\{\left\|\nabla_{\frac{\partial}{\partial x}}w\right\|^2 + \left\|\nabla_{\frac{\partial}{\partial y}}w\right\|^2\right\} dx\, dy - \sum_j \int_M g(x)\left(R\left(\frac{\partial S}{\partial x^j}, w\right)w, \frac{\partial S}{\partial x^j}\right) dx\, dy$$

we arrive at the conclusion of Theorem E.1.

Remark E.1 *The similarity between the formula (E.1) for the Levi-form in Abresch-Fischer coordinates and formula (6.8) is no accident. It is a consequence of the fact that Abresch-Fischer coordinates almost satisfy the condition of Deligne, Griffiths, Morgan and Sullivan (cf. theorem 6.2.1). See remark 6.2.2.*

F Riemann-Roch and the Dimension of Teichmüller Space

In this chapter we briefly give the background material from Riemann surface theory that enables one to state the famous theorem of Riemann-Roch and to compute the dimension of the space of holomorphic quadratic differentials on a Riemann surface.

Let (M, c) be a surface of genus greater than one with an associated complex structure c. For ease of exposition we shall suppress the c from the notation (M, c) for the rest of the chapter. Given such a structure c we clearly have the notion of a meromorphic function on M. From the maximum modulus principle it follows that the only holomorphic functions on M are constants. Nevertheless, we can have holomorphic differentials.

Let ω be a complex valued one form on M; i.e. for $x \in M$, $\omega(x) : T_x M \to \mathbb{C}$ is linear over the real. In a local coordinate chart with local variables designated by x and y we can represent ω by

$$\omega = P \, dx + Q \, dy + i(\tilde{P} \, dx + \tilde{Q} \, dy)$$

We say that ω is *holomorphic* differential if "locally" ω can be written as

$$\omega(z) = f'(z)dz$$

$dz = dx + i \, dy$, and where f is a holomorphic function. It is a well known fact (thm 10.3 of [104]) that the complex dimension of the vector space of holomorphic differentials is equal to the genus of M. One also has the obvious notion of a *meromorphic* differential by requiring f to be meromorphic. A complex 1-form on M which is either holomorphic or meromorphic is called an *abelian* differential.

It is easy to check that the order of a zero or a pole of either a meromorphic function or an abelian differential is well-defined. We are interested in specifying to some extent the location and orders of poles of both meromorphic functions and abelian differentials on M.

Let $P_1, P_2, \ldots, P_n$ be points on M and $\alpha_1, \alpha_2, \ldots, \alpha_n$ be integers. The symbol

$$\mathbf{a} = P_1^{\alpha_1} P_2^{\alpha_2} \cdots P_n^{\alpha_n}$$

is called a *divisor*. The integer α_k is called the *order* at P_k. By the *degree* $d[\mathbf{a}]$ of a divisor $\mathbf{a}$ we mean the sum

$$d[\alpha] = \Sigma \alpha_k \ .$$

If f is a meromorphic function not identically zero on M and $\omega \not\equiv 0$ is an abelian differential we define the divisors (f) and (ω) of f and ω by

$$(f) = P_1^{\alpha_1} \cdots P_k^{\alpha_k} Q_1^{-\beta_1} \cdots Q_\ell^{-\beta_\ell}$$

where the zeros of f are $P_1, \ldots, P_k$ with orders $\alpha_1, \ldots, \alpha_k$, all $\alpha_i \geq 0$, and the poles of f are $Q_1, \ldots, Q_\ell$ with orders $\beta_1, \ldots, \beta_\ell$, all $\beta_i \geq 0$ and

$$(\omega) = \tilde{P}_1^{\tilde{\alpha}_1} \cdots \tilde{P}_r^{\tilde{\alpha}_r} \tilde{Q}_1^{-\tilde{\beta}_s} \cdots \tilde{Q}_s^{-\tilde{\beta}_s}$$

where $\tilde{P}_1, \ldots, \tilde{P}_r$ are the zeros of ω of orders $\tilde{\alpha}_1, \ldots, \tilde{\alpha}_r$ and $\tilde{Q}_1, \ldots, \tilde{Q}_s$ are the poles of ω of orders $\tilde{\beta}_1, \ldots, \tilde{\beta}_s$.

Since for a meromorphic function, the sum of the orders of its zeros is equal to the sum of the orders of its poles it follows that $d(f) = 0$ for any f.

The following is a basic result in Riemann surface theory.

Theorem F.1 *If ω is an abelian differential then $d(\omega) = 2 \ genus(M) - 2$.*

Note that $d(\omega) > 0$ in our case where genus$(M) > 1$. Consequently every abelian differential must have a zero.

A divisor $\mathbf{a} = P_1^{\alpha_1} \cdots P_k^{\alpha_k}$ is called integral if $\alpha_j \geq 0$ for all j. If $\mathbf{b} = Q_1^{\beta_1} \cdots Q_1^{\beta_1}$ then the quotient divisor $\mathbf{a}/\mathbf{b}$ is defined by

$$\mathbf{a}/\mathbf{b} = P_1^{\alpha_1} \cdots P_k^{\alpha_k} Q_1^{-\beta_1} \cdots Q_1^{-\beta_1} \ .$$

By $\frac{1}{\mathbf{a}}$ we mean the divisor $P_1^{-\alpha_1} \cdots P_k^{-\alpha_k}$. If $\mathbf{a}/\mathbf{b}$ is integral we say that $\mathbf{b}$ divides $\mathbf{a}$ or that $\mathbf{a}$ is a multiple of $\mathbf{b}$.

Define by $L(\mathbf{a})$ the vector space of meromorphic functions on M whose divisors are integral multiples of $\mathbf{a}$ and by $\Omega(\mathbf{a})$ the vector space of abelian differentials whose divisors are integral multiples of $\mathbf{a}$.

A beautiful relationship between the dimensions of these vector spaces over the complexes is given by

Theorem F.2 (Riemann-Roch)

$$\dim L\left(\frac{1}{\mathbf{a}}\right) = \dim \Omega(\mathbf{a}) + d[\mathbf{a}] - \text{genus}(M) + 1 \ .$$

For a proof the reader may consult any number of elementary texts on Riemann surfaces, see for example [104], [37].

In addition to permitting us to speak about meromorphic functions and abelian differentials a complex structure on M allows us to speak about holomorphic or meromorphic quadratic differentials.

A complex valued quadratic differential on M is a complex valued symmetric 0-2 tensor Q. Thus for each $q \in M$,

$$Q(q) : T_q M \times T_q M \rightarrow \mathbb{C}$$

is bilinear and symmetric. Locally Q can be expressed as

$$Q = E \ dx^2 + F \ dx \ dy + G \ dy^2 + i(\tilde{E} \ dx^2 + \tilde{F} \ dx \ dy + \tilde{G} \ dy^2) \ .$$

Q is said to be *holomorphic* if it can be expressed in a local complex coordinate system as

$$Q(z) = \varphi(z) dz^2 \ .$$

with φ holomorphic. Let $\mathcal{Q}(M)$ denote the complex linear space of holomorphic quadratic differentials on M. The following theorem on the dimension of $\mathcal{Q}(M)$ over the complexness is the principal result we will need from elementary Riemann surface theory.

Theorem F.3 $\dim_{\mathbb{C}} \mathcal{Q}(M) = 3\text{genus}(M) - 3$. *Therefore* $\dim_{\mathbb{R}} \mathcal{Q}(M) = 6\text{genus}(M) - 6$.

PROOF: Let $\omega_0(z) dz^2$ be a holomorphic quadratic differential, say for example the product of two holomorphic abelian differentials. Then it follows from F.1 that if $\mathbf{a}_0$ is the divisor of ω_0 then $d[\mathbf{a}_0] = 4\text{genus}(M) - 4$. If $\omega(z) dz^2$ is any other holomorphic quadratic differential then $\omega(z)dz^2/\omega_0(z)dz^2$ is a meromorphic function. If **a** denotes the divisor of ω and since $\mathbf{aa_0^{-1}}$ is the divisor of a meromorphic function it follows that $d[\mathbf{a}] = d[\mathbf{a_0}] = 4\,\text{genus}(M) - 4$.

For arbitrary ω let $f_\omega(z) = \omega(z)dz^2/\omega_0(z)dz^2$. Then $\omega = f_\omega \cdot \omega_0$ where (f_ω) is an integral multiple of $\mathbf{a_0^{-1}}$. It therefore follows that the elements of $\mathcal{Q}(M)$ are in one to one correspondence with $L(\mathbf{a_0^{-1}})$. By the Riemann-Roch theorem

$$\dim L(\mathbf{a_0^{-1}}) = \dim \Omega(\mathbf{a_0}) + d[\mathbf{a_0}] - \text{genus}(M) + 1 \ .$$

If $\tau = \varphi(z)dz \in \Omega(\mathbf{a_0})$ is non-zero, $d(\tau) \geq d[\mathbf{a_0}] = 4\,\text{genus}(M) - 4$. But by F.1, $d(\tau) = 2\text{genus}(M) - 2$ which is impossible. Thus $\dim \Omega(\mathbf{a_0}) = 0$. Hence

$$\dim_{\mathbb{C}} \mathcal{Q}(M) = \dim L(\mathbf{a_0^{-1}}) = 4\,\text{genus}(M) - 4 - \text{genus}(M) + 1 = 3\,\text{genus}(M) - 3\ .$$

Bibliography

[1] **W. Abikoff:** The real analytic theory of Teichmüller space, Springer Lecture Notes in Math. 820, 1980

[2] **R. Abraham, J. Marsden, T. Ratiu:** Manifolds, Tensor Analysis, and Applications, Addison-Wesley, 1983 .

[3] **R. Adams:** Sobolev Spaces, Springer 1985.

[4] **L.V. Ahlfors:** On quasiconformal mappings, J. d'Analyse Math. **3** (1953/54), 1-58, 207-208.

[5] **L.V. Ahlfors:** Curvature properties of Teichmüller's space, J. d'Analyse Math. **9** (1961), 161-176.

[6] **L.V. Ahlfors:** The complex analytic structure on the space of closed Riemann surfaces, in: Analytic functions, Princeton UPr 1960.

[7] **L.V. Ahlfors:** Lectures on quasiconformal mappings, Van Nostrand-Reinhold, Princeton NJ, 1966

[8] **L.V. Ahlfors, L. Sario:** Riemann surfaces, Princeton UPr 1960

[9] **R. Baer:** Isotopie von Kurven auf orientierbaren geschlossenen Flächen, Crelle J. f. reine u. angewandte Math. **159** (1928), 101-116.

[10] **H. Behnke, F. Sommer:** Theorie der analytischen Funktionen einer komplexen Veränderlichen, Springer 1955.

[11] **L. Bers:** Quasiconformal mappings and Teichmüller theory, in: Analytic functions, 89-120, Princeton UP 1960.

[12] **L.Bers:** Simultaneous Uniformization, Bull. AMS **66** (1960), 94-97.

[13] **L. Bers:** Spaces of Riemann Surfaces as Bounded Domains, Bull. AMS, **66** (1960), 98-103; a correction to this paper: Bull. AMS **67** (1961), 465-466.

[14] **L. Bers, L. Ehrenpreis:** Holomorphic convexity of Teichmüller Spaces, Bull. AMS **70** (1964), 761-764.

[15] **J. Cheeger, D. Ebin:** Comparison Theorems in Riemannian Geometry, North Holland Publ., Amsterdam 1975

[16] **P. Deligne, P. Griffiths, J. Morgan, D. Sullivan:** Real Homotopy Theory of Kähler Manifolds, Inv. math. **29** (1975), 245-274.

[17] **J. Dieudonné:** Foundations of Modern Analysis, Academic Press, 1960.

[18] **D. Ebin:** Manifolds of Riemannian Metrics, AMS Proc. on Global Analysis, Berkeley 1968

[19] **C. Earle:** On holomorphic cross sections in Teichmüller spaces, Duke Math. Journal **36** (1969), 409-415.

[20] **C. Earle, J. Eells:** Deformations of Riemannian Surfaces, Lectures in Modern Analysis and Appl. **103** (1969), 122-149, Springer

[21] **C. Earle, J. Eells:** The diffeomorphism group of a compact Riemann surface, Bull. Amer. Math. Soc. **73** (1967), 557-559.

[22] **C. Earle, J. Eells:** A fibre bundle description of Teichmüller theory, J. Diff. Geom. **3** (1969), 19-43.

[23] **B. Eckmann, H. Müller:** Poincaré duality groups of dimension two, Comment. Math. Helvetici **55** (1980), 510-520.

[24] **B. Eckmann, H. Müller:** Plane motion groups and virtual Poincaré duality of dimension 2, Inventiones Math. **69** (1982), 293-310.

[25] **B. Eckmann:** Poincaré duality groups of dimension 2 are surface groups (survey), Ann. Math. Studies 1986, 35-51. Princeton UPr

[26] **J. Eells, L. Lemaire:** A report on harmonic maps, Bull. London Math. Soc. **10** (1978), 1-68.

[27] **J. Eells, L. Lemaire:** Another report on harmonic maps, Bull. London Math. Soc. **20** (1988), 385-524.

[28] **J. Eells, J.H. Sampson:** Harmonic mappings of Riemannian Manifolds, American Journal of Mathematics **86** (1964), 109-160.

[29] **L.P. Eisenhart:** Riemannian geometry, Princeton UP 1966.

[30] **W. Fenchel:** Estensioni di gruppi discontinui e trasformazioni periodiche delle superficie, Atti della Accademia Nazionale dei Lincei, Classe di scienze fisiche, matematiche e naturali, serie 8; **5** (1948), 326-329.

[31] **W. Fenchel:** Bemærkninger om endelige grupper af afbildningsklasser, Matematisk Tidsskrift B, 1950, 90-95.

[32] **A.E. Fischer, J.E. Marsden:** Deformation of Scalar Curvature, Duke Math. Jour. **42** (1975), 519-547.

[33] **A.E. Fischer, A.J. Tromba:** On the Weil-Petersson metric on Teichmüller space, Trans. AMS **284** (1984), 319-335.

[34] **A.E. Fischer, A.J. Tromba:** Almost complex principle bundles and the complex structure on Teichmüller space, Crelles J. Band **252**, 151-160.

[35] **A.E. Fischer, A.J. Tromba:** On a purely Riemannian proof of the structure and dimension of the unramified moduli space of a compact Riemann surface, Math. Ann. **267** (1984), 311-345.

[36] **A.E. Fischer, A.J. Tromba:** A new proof that Teichmüller space is a cell, Trans. AMS vol. **303**, No. 1 Sept. (1987), 257-262.

[37] **O. Forster:** Riemannsche Flächen, Springer Heidelberger Taschenbücher 184.

[38] **R. Fricke, F. Klein:** Vorlesungen über die Theorie der automorphen Funktionen, Teubner, Leipzig 1926.

[39] **A. Friedman:** Partial Differential Equations, Robert E. Krieger Publishing Co., Huntington NY, 1976

[40] **F.P. Gardiner:** Teichmüller Theory and Quadratic Differentials, Wiley 1987

[41] **M. Giaquinta, S. Hildebrandt:** A priori estimates for harmonic mappings, Crelles Jour. f. reine u. angew. Math. **336** (1982), 124-164.

[42] **D. Gilbarg, N.S. Trudinger:** Elliptic Partial Differential Equations of Second Order, Grundlehren vol. **224** (1983), Springer, Berlin (2nd edn.).

[43] **A. Gramain:** Topologie des surfaces, Presses universitaires de France, 1971.

[44] **H. Grauert, R. Remmert:** Theorie der Steinschen Räume, Springer Grundlehren 227, (1977); English translation: Theory of Stein Spaces, Springer Grundlehren 236, (1979)

[45] **V. Guillemin, A. Pollack:** Differential Topology, Prentice Hall 1974.

[46] **R.C. Gunning:** Introduction to Holomorphic Functions of Several Complex Variables, 3 vols, Wadsworth&Brooks/Cole, Math. Series, Belmont, California, 1990

[47] **R.C. Gunning, H. Rossi:** Analytic Functions of Several Complex Variables, Prentice Hall, Eaglewood Cliffs, N.J., 1965

[48] **N. Halpern:** A proof of the collar lemma, Bull. London Math. Soc. **13** (1981), 141-144.

[49] **R.S. Hamilton:** Harmonic maps of manifolds with boundary, Springer Lecture Notes in Math. #471, 1975

[50] **P. Hartmann:** On homotopic harmonic maps, Can. J. Math. **19** (1967), 673-687.

[51] **W.T. Harvey:** Discrete Groups and Automorphic Functions, Academic Press, 1977.

[52] **E. Heinz:** On certain non-linear elliptic differential equations and univalent mappings, Journal d'analyse **5** (1956/57), 197-272.

[53] **S. Hildebrandt:** Harmonic mappings of Riemannian manifolds; pp 1-117 in: Harmonic Mappings and minimal Immersions, Springer Lecture Notes #1161, 1985

[54] **S. Hildebrandt, H. Kaul, K. Widman:** Harmonic mappings into Riemannian manifolds with non-positive sectional curvature, Math. Scand. **37** (1975), 257-263.

[55] **S. Hildebrandt, H. Kaul, K. Widman:** Dirichlet's boundary value problem for harmonic mappings of Riemannian manifolds, Math. Zeitschrift **147** (1976), 225-236.

[56] **S. Hildebrandt, H. Kaul, K. Widman:** An existence theorem for harmonic mappings of Riemannian manifolds, Acta Math. **138** (1977), 1-16.

[57] **J. Jost:** Two Dimensional Geometric Variational Problems, Pure and Applied Mathematics Series, Wiley, 1991

[58] **J. Jost, R. Schoen:** On the existence of harmonic diffeomorphisms between surfaces, Inv. Math. **66** (1982), 353-359.

[59] **L. Keen:** Collars on Riemann surfaces, Ann. Math. Studies **79** (1974), 263-268.

[60] **S.P. Kerckhoff:** The Nielsen Realization Problem, Bull. AMS **2** (1980), 452-454. (announcement)
Annals of Math., ser.2, **117** (1983), 235-265. (details)

[61] **S. Kravetz:** On the geometry of Teichmüller spaces and the structure of their modular group, Ann. Acad. Fennicae, Ser.A VI **278** (1959), 1-35.

[62] **A. Kufner, O. John, S. Fučik:** Function Spaces, Noordhof Int. Publ., 1977.

[63] **S. Lang:** Analysis I+II, Addison-Wesley, 1968/69.

[64] **S. Lang:** Introduction to Differentiable Manifolds, Interscience, New York, 1962

[65] **O. Lehto:** Univalent functions and Teichmüller spaces, Springer Graduate Text #109, 1987

[66] **L. Lemaire:** Applications harmoniques de surfaces Riemanniennes, J. Diff. Geom. **13** (1978), 51-78.

[67] **L. Lemaire:** Boundary value problems for harmonic and minimal maps of surfaces into manifolds, Ann. Scuola Norm. Sup. Pisa, classe di sci. mat., ser. 4, **9** (1982), 91-103.

[68] **E.E. Levi:** Studii sui punti singolari essenziali delle funzioni analitiche di due o più variabili complesse, Annali di matematica pura ed applicata, ser. 3, **17** (1910), 61-87.

[69] **A. Lichnerowicz:** Propagateurs et commutateurs en relativité générale, Publ. math., IHES, **10** (1961).

[70] **A.M. Macbeath:** On a theorem by J. Nielsen, Quarterly J. Math. Oxford, Ser.2, **13** (1962), 235-236.

[71] **J. McCleary:** User's Guide to Spectral Sequences, Publish or Perish, 1985

[72] **W. Mangler:** Die Klassen von topologischen Abbildungen einer geschlossenen Fläche auf sich, Math. Zeitschrift **44** (1939), 541-554.

[73] **W.S. Massey:** Algebraic Topology: An Introduction, Harbrace 1967 or Springer 1977

[74] **H. Masur:** The curvature of Teichmüller space, Springer Lecture Notes in Math. #400, 1974

[75] **V.G. Maz'ja:** Sobolev Spaces, Springer 1985.

[76] **J. Milnor:** Morse Theory, Princeton UP, 1963

[77] **C.W. Misner, K.S. Thorne, J.A. Wheeler:** Gravitation, Freeman, 1973.

[78] **Ch.B. Morrey:** Multiple Integrals in the Calculus of Variations, Springer Grund-lehren 130, 1966

[79] **D. Mumford:** A remark on Mahler's compactness theorem, Proc. Am. Math. Soc. **28** (1971), 288-294.

[80] **A. Newlander, L. Nirenberg:** Complex analytic coordinates in almost complex manifolds, Ann. Math. **65** (1957), 391-404.

[81] **J. Nielsen:** Untersuchungen zur Topologie der geschlossenen zweiseitigen Flächen I, Acta Math. **50** (1927), 189-358.

[82] **J. Nielsen:** Untersuchungen zur Topologie der geschlossenen zweiseitigen Flächen II, Acta Math. **53** (1929), 1-76.

[83] **J. Nielsen:** Untersuchungen zur Topologie der geschlossenen zweiseitigen Flächen III, Acta Math. **58** (1932), 87-167.

[84] **J. Nielsen:** Die Struktur periodischer Transformationen von Flächen, Matematisk-fysiske meddelelser **15** (1937), 1-77, Danske videnskabernes selskab.

[85] **J. Nielsen:** Über Gruppen linearer Transformationen, Festschrift Teil II, Mitteilun-gen Math. Ges. Hamburg **8** (1940), 82-104.

[86] **J. Nielsen:** Abbildungsklassen endlicher Ordnung, Acta Math. **75** (1942), 23-115.

[87] **B. O'Neill:** The fundamental equations of a submersion, Michigan Math. Journal **13** (1966), 459-469.

[88] **K. Oka:** Sur les fonctions analytiques de plusieurs variables, Iwanami Shoten, Tokyo 1961

[89] **R. Palais:** Morse Theory on Hilbert Manifolds, Topology **2** (1963), 299-340.

[90] **R. Palais:** Foundations of Global Non-Linear Analysis, Benjamin, New York, 1968

[91] **M.H. Protter, H.F. Weinberger:** Maximum Principles in Differential Equations, Prentice Hall, 1967

[92] **B. Randol:** Cylinders in Riemann surfaces, Commentarii Math. Helvetici **54** (1979), 1-5.

[93] **B. Riemann:** Theorie der Abelschen Funktionen, Borchardt's Journal f. reine u. angew. Mathematik **54** (1857); reprinted in: Gesammelte Mathematische Werke, Wiss. Nachlass und Nachträge, Collected Papers, Springer and Teubner-Leipzig, 1990

[94] **H. Royden:** Automorphisms and isometries of Teichmüller space, Ann. Math. Studies **66** (1966), 369-383.

[95] **J. Sacks, K. Uhlenbeck:** The existence of minimal immersions of 2-spheres, Annals of Math. **113** (1981), 1-24.

[96] **J.H. Sampson:** Some properties and applications of harmonic mappings, Annales Sci. Ecole Normale Supérieure **11** (1978), 211-228.

[97] **R. Schoen, S.T. Yau:** On univalent harmonic maps between surfaces, Inv. Math. **44** (1978), 265-278.

[98] **H. Seifert:** Bemerkung zur stetigen Abbildung von Flächen, Abhandlungen des Math. Seminars, Univ. Hamburg **12** (1938), 29-37.

[99] **Y.T. Siu:** Curvature of the Weil-Petersson metric in the moduli space of compact Kähler-Einstein manifolds of negative first Chern class, in K. Diederich (ed): Aspects of Mathematics, vol 9, pp 261-298, Vieweg

[100] **S. Smale:** Morse theory and a non-linear generalization of the Dirichlet problem, Annals of Math., ser.2, **80** (1964), 382-396.

[101] **P.A. Smith:** A theorem on fixed points for periodic transformations, Annals of Math. **35** (1934), 572-578.

[102] **E.H. Spanier:** Algebraic Topology, McGraw-Hill 1966, now Springer

[103] **M. Spivak:** A Comprehensive Introduction to Differential Geometry, 5 vols, Publish or Perish Press

[104] **G. Springer:** Introduction to Riemann Surfaces, Addison Wesley, 1957

[105] **N. Steenrod:** The Topology of Fibre Bundles, Princeton UPr 1951

[106] **O. Teichmüller:** Extremale quasikonforme Abbildungen und quadratische Differentiale, Abh. Preuss. Akad. Wiss., Math.-Naturw. Klasse 4 (1943), 1-197 (also in: [108, pp. 335-531]).

[107] **O. Teichmüller:** Veränderliche Riemannsche Flächen, Deutsche Math. **7** (1944), 344-359 (also in: [108, pp. 712-727]).

[108] **O. Teichmüller:** Gesammelte Abhandlungen - Collected Papers, ed. L. Ahlfors, F. Gehring, Springer, Berlin 1982.

[109] **F. Tomi:** Über elliptische Differentialgleichungen 4. Ordnung mit einer starken Nichtlinearität, Nachrichten der Akademie der Wissenschaften zu Göttingen Nr. **3** (1976), 1-10.

[110] **F. Tomi, A.J. Tromba:** Existence theorems for minimal surfaces of non-zero genus spanning a contour, Mem. AMS, vol **71**, number 382, Providence, RI, 1988

[111] **A.J. Tromba:** A General Approach to Morse Theory, Journal of Differential Geometry **12** (1977), 47-85.

[112] **A.J. Tromba:** On a natural algebraic affine connection on the space of almost complex structures and the curvature of Teichmüller space with respect to its Weil-Petersson metric, Manuscripta Math. **56**, Fas. 4 (1986), 475-497.

[113] **A.J. Tromba:** On an energy function for the Weil-Petersson metric, Manuscripta Math. **59** (1987), 249-266.

[114] **A.J. Tromba:** Dirichlet's Energy on Teichmüller's Moduli Space is Strictly Pluri-Subharmonic, SFB 256 Bonn preprint # 207 (1992).

[115] **A.J. Tromba:** On the Levi-form for Dirichlet's energy on Teichmüller space, SFB 256 Bonn preprint # 206 (1992).

[116] **A.J. Tromba:** Dirichlet's energy and the Nielsen Realization Problem, SFB 256 Bonn preprint # 208 (1992).

[117] **A.J. Tromba:** A new proof that Teichmüller space is a cell, Trans. Am. Math. Soc. **303** (1990), 257-262.

[118] **R. Schoen, S.T. Yau:** On univalent harmonic maps between surfaces, Inventiones mathematicae, **44** (1978), 265-278.

[119] **K. Uhlenbeck:** Harmonic maps; A direct method in the calculus of variations, Bull. AMS **76** (1970), 1082-1087.

[120] **K. Uhlenbeck:** Morse theory by perturbation methods with applications to harmonic maps, Transactions AMS **276** (1981), 569-583.

[121] **B. White:** Mappings that minimize area in their homotopy classes, Jour. Diff. Geometry, **20** (1984), 433-446.

[122] **G.W. Whitehead:** Elements of Homotopy Theory, Springer, 1978, (Graduate Texts in Math. # 61)

[123] **M. Wolf:** The Teichmüller Theory of Harmonic Maps, J. Diff. Geometry **29** (1989), 449-479.

[124] **S. Wolpert:** Geodesic length functionals and the Nielsen problem, J. Diff. Geometry **25** (1987), 275-295.

[125] **S. Wolpert:** Chern forms and the Riemann tensor for the moduli space of curves, Inv. Math. **85** (1986), 119-145.

[126] **S. Wolpert:** Noncompleteness of the Weil-Petersson metric for Teichmüller space, Pacific J. Math. **61** (1975), 573-577.

[127] **H. Zieschang:** Über Automorphismen ebener diskontinuierlicher Gruppen, Math. Annalen **166** (1966), 148-167.

[128] **H. Zieschang:** On extensions of fundamental groups of surfaces and related groups, Bull. Amer. Math. Soc. **77** (1971), 1116-1119.

[129] **H. Zieschang:** On triangle groups, Russian Mathematical surveys, **31**,5 (1976), 226-233. Translated from: **Х. Цишанг:** О треугольных группах, Успехи мат. наук **31**,5 (1976), 177-183.

[130] **H. Zieschang:** Finite Groups of Mapping Classes of Surfaces. Lecture Notes in Mathematics 875, Springer Verlag 1981.

[131] **H. Zieschang, E. Vogt, H-D. Coldewey:** Surfaces and Planar Discontinuous Groups. Lecture Notes in Mathematics 835, Springer Verlag 1980.

We include a list of some notations here. It does not contain every single item, but should include the basic symbols and where they occur. An asterisk in the column "where?" means that the notation is used throughout the book and no reference for a definition is needed. Otherwise, main occurrences and places of definitions are indicated.

symbol	what?	where?
a,b	divisors	app. F
$\mathcal{A}, \mathcal{A}^a$	manifold of almost complex structures	def. 1.1.1
$c_1, C_1, \cdots$	constants	*
c	a complex structure on M	def. 0.2, sec. 1.1
C	the set of complex structures on M	p. 10
C^1, C^k, C^∞	$1, k, \infty$ times continuously differentiable	*
$\mathbb{C}$	the set of complex numbers	*
$d[\mathbf{a}]$	degree of a divisor	app. F
$\mathcal{D}$	the set of C^∞ diffeomorphisms of M	p. 10
$\mathcal{D}_0$	— " — homotopic to the identity	p. 10
$Df(g)h$	derivative of the function f at g in direction h	*
$DX(Y)$	for vector fields X, Y the function $p \mapsto DX(p)Y(p)$, extrinsic	ch. 5
$D_Y X$	— " —	sec. 5.3
$\frac{D}{\partial t}$	covariant derivative, extrinsic	sec. 2.1, 3.1, app. B
e	energy density	sec. 3.1, app. B
E, E_g	Energy functional, Dirichlet's energy	ch. 3
$\hat{E}, \bar{E}$	Dirichlet's energy on $\mathcal{M}, \mathcal{T}(M)$	sec. 3.1
E, E_G	Wolf's form of Dirichlet's energy functional	sec. 6.3
$\bar{\mathsf{E}}$	Wolf's form of Dirichlet's energy on $\mathcal{T}(M)$	sec. 6.3
$\mathcal{E}_u$	the differential operator defined in	(B.5), app. B
f	usually an element of $\mathcal{D}$	*
$\mathcal{F}$	a real valued function on a complex manifold	ch. 6
g, g_0	a metric on M	*
G, G_Σ	a metric / induced metric on Σ	*/thm. 5.1.1
$\mathcal{G}$	a (Hilbert) Lie group	chs. 4,5
h	a 0-2 tensor field, tangent vector to $\mathcal{M}, \mathcal{M}_{-1}, \mathcal{T}(M)$	*

symbol	what?	where?
^{H}	horizontal component	p. 99
H	a 1-1 tensor field on M	*
$\mathbb{H}$	the upper half plane (hyperbolic plane)	*
$H^\bullet(M)$	functions on M of Sobolev class $H^\bullet$	ch. 0
$\mathcal{H}^\bullet(T_q^p M)$	tensor fields on M of class $H^\bullet$, p times contravariant and q times covariant	ch. 0
I	unit matrix or identity map on some vector space	*
id	identity map	*
J	an almost complex structure on M, i.e. an element of $\mathcal{A}^\bullet$	def. 1.1.1
J	Jacobian determinant	p. 180
K	a 1-1 tensor field on M	*
$\mathcal{K}$	sectional curvature	sec. 5.4
L_X	Lie derivative in direction X	(1.3)
$L_k^p(M, \mathbb{R}^d)$	Sobolev space	p. 159
$\mathcal{L}$	$\Delta - id$	p. 115
M	Riemann surface of genus ≥ 2	*
$\mathcal{M}, \mathcal{M}^\bullet$	manifold of C^∞ or $H^\bullet$ metrics on M	sec. 1.2
$\mathcal{M}_{-1}, \mathcal{M}^\bullet_{-1}$	manifold of C^∞ or $H^\bullet$ metrics on M with scalar curvature -1	secs. 1.2, 1.6
N	some manifold	*
$\mathbf{N}$	Nijenhuis tensor	def. 4.1.3
$O_f, \hat{O}_f$	the action by f on some space	*
p	a point on a manifold; or a positive function in the context of conformal coordinates	*
P	total space of a principal bundle	sec. 4.1
$\mathcal{P}, \mathcal{P}^\bullet$	positive functions on M	sec. 1.3
$\mathcal{Q}(M)$	space of holomorphic quadratic differentials on M	app. F
pr	a projection map	*

symbol	what?	where?
R	scalar curvature	p. 24
$\mathcal{R}, \hat{\mathcal{R}}$	Riemann curvature tensor	sec. 5.4
$\mathcal{R}(M)$	Riemann moduli space	def. 0.5
$\mathbb{R}$	the set of real numbers	*
S	a harmonic map	p. 64
S_2	symmetric 0-2 tensors	ch. 0
S_2^{TT}	space of transverse (=divergence free) traceless 0-2 tensors, i.e. the tangent space to $\mathcal{T}(M) \cong \mathcal{M}_{-1}/\mathcal{D}_0$	p. 45
$\mathcal{S}$	"slice"	pp. 47–57
$SL_2(\mathbb{R})$	the group of 2×2 matrices of determinant 1, two sheeted cover of automorphisms of upper half plane	*
^{T}	trace free part	p. 19, sec. 1.3
^{TT}	transverse (=divergence free) traceless	p. 45
$T_x M$	tangent space at x to M	*
$\mathcal{T}(M)$	Teichmüller space	def. 0.6
tr	trace of a 1-1 tensor or linear map	*
tr_g	trace with respect to the metric g	*
u	a function from M to M	app. B
U	an open set; domain of coordinate function or of its inverse	*
^{V}	vertical component	p. 99
$\mathcal{X}(M)$	space of C^∞ vector fields on M	*
X, Y, Z	vector fields	*
z	complex coordinate on M	*

symbol	what?	where?
α_J	the map $X \mapsto L_X J$	sec. 1.4
$\beta(u)$	the vector field over u defined in	(B.8)
Γ_{ij}^k	Christoffel symbol	*
δ_{ij}	Kronecker delta	*
δ_g	divergence	sec. 1.4
$\Lambda(f)$	Lefshetz number of f	p. 39
λ	conformal factor in the context of conformal coordinates	*
λ, μ	given $J \in \mathcal{A}$, certain functions on M	sec. 5.4
π	projection map in a bundle	*
Π	projection map in a linear space	*
σ, σ_{EE}	a section in a bundle, Earle-Eells section	sec. 3.4
$\sigma(t)$	a geodesic	sec. 2.1
Σ	base space of a principal bundle	sec. 4.1
$\Phi, \hat{\Phi}$	almost complex structure on $\mathcal{A}$, $\mathcal{T}(M)$	ch. 5
φ, ψ	coordinate map or its inverse	*
$\chi(M)$	Euler characteristics of M	*
Ω	lattice in $\mathbb{C}$	p. 8
Ω	Kähler form on $\mathcal{A}$	sec. 5.1
Ω_Σ, Ω	Kähler form on Σ resp. $\mathcal{T}(M)$	sec. 5.1
$\langle \cdot, \cdot \rangle$	inner product (specified in the context)	*
$\langle \cdot, \cdot \rangle_{WP}$	Weil-Petersson metric	(2.7)
$\langle\langle \cdot, \cdot \rangle\rangle$	L^2-inner product for 1-1 tensors and on $\mathcal{M}$	pp. 19,56
$\langle\langle\langle \cdot, \cdot \rangle\rangle\rangle$	alternative L^2-inner product on $\mathcal{M}_{-1}$	sec. 2.5,2.6
∇_g	gradient with respect to g	*
$\nabla, \hat{\nabla}$	connection, Levi-Cività connection	sec. 5.4
Δ_g	Laplace-Beltrami operator	sec. 1.4
Δ	non-linear Laplacian of a map $M \to M$ or linear Laplacian of a vector field over u	p. 160
$:=, =:$	equal by definition; the colon is on the side being defined	*
$\to:$	converges to something, the name of the limit is defined	*
$*\omega$	the Hodge dual of a differential form ω	ch. 5,6, app. E
$(\cdot)$	L^2-inner product density of 0-2 tensors (in $S_2^{TT} \cong T\mathcal{M}_{-1}$)	p. 72, ch. 6, app. E

The Maps Used in the Construction of Teichmüller space

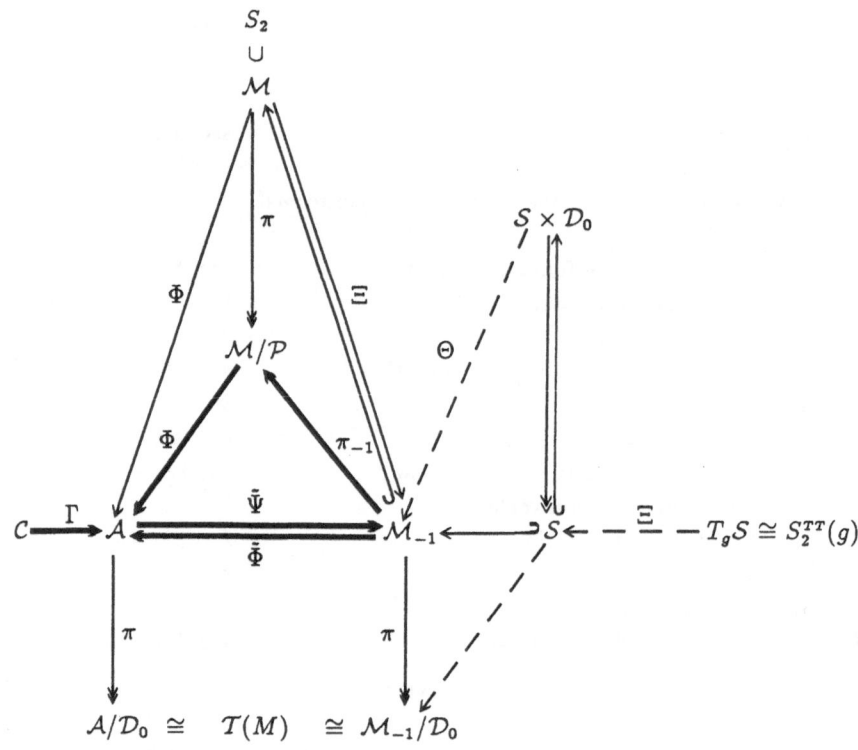

───────▶ The fat arrows denote $\mathcal{D}$-equivariant diffeomorphisms or $\mathcal{D}$-equivariant bijective maps

─ ─ ─ → The dashed arrows denote maps defined only on a neighbourhood of some point(s) which are diffeomorphisms from such a neighbourhood to their image.

─────≫ surjective maps

⊂────→ natural inclusions

The diagram is commutative.

Index